# Studies in Fuzziness and Soft Computing    289

**Editor-in-Chief**

Prof. Janusz Kacprzyk
Systems Research Institute
Polish Academy of Sciences
ul. Newelska 6
01-447 Warsaw
Poland
E-mail: kacprzyk@ibspan.waw.pl

For further volumes:
http://www.springer.com/series/2941

Kofi Kissi Dompere

# The Theory of the Knowledge Square: The Fuzzy Rational Foundations of the Knowledge-Production Systems

 Springer

*Author*
Kofi Kissi Dompere
Department of Economics
Howard University
Washington, D.C.
USA

ISSN 1434-9922                          e-ISSN 1860-0808
ISBN 978-3-642-44577-4           ISBN 978-3-642-31119-2  (eBook)
DOI 10.1007/978-3-642-31119-2
Springer Heidelberg New York Dordrecht London

Printed on acid-free paper

Springer is part of Springer Science+Business Media (www.springer.com)

# Dedication

*To all scholars, researchers, supporters and authors in the reference list whose efforts and hard works have made the work in this monograph possible, less tasking and more enjoyable.*

## IN APPRECIATION

Of the works of Springer, especially Professor Kacprzky and the editorial team of the series of Fuzziness and Soft Computing. This appreciation is also extende to Professor Zimamann, the editorial board and the refereeing network of the International Journal of Fuzzy Set and System, Busefal, Journal of Fuzzy Mathematics, International Journal of Fuzzy Economics, IEEE Systems Man and Cybenetics, Automatic Control System, Multi-value Logics for promoting research on the fuzzy phenomenon.

# Acknowledgements

The theory of knowing, also characterized as the theory of knowledge, irrespective of how it is conceived and interpreted, affects all areas of human action and the practice of results of thinking that may guide the development of social and physical technologies in the social setups in time and over time. The understanding the path of knowing is very important to the success of the knowing process. There are many epistemic paths to the development of the knowing and knowledge-production processes. Not all the available paths will lead to an efficient utilization of available resource and cognitive time. One path may be conceived from the viewpoint of rational construct while another may be viewed as accident of knowing. Any of these paths to knowing, therefore, is about the discovery of *what there is* (or *what ought to be*) in human action and explanation (or implementation) of the behavior of how *what there is* (or *what out to be*) manifests itself in information-knowledge structures through the epistemic decision-choice processes. The theory of the knowledge square is about the analysis and discovery of an efficient path of knowing of the elements in the universal object set relative to social and natural states and processes under the conditions of language vagueness, human cognitive limitation and epistemic ambiguities that lead to exactness in knowing in an inexact information environment. In this frame, rationality, therefore, is about the discovery of an efficient path of intelligence in human action, and the explanation as to how this intelligence allows the universal object set to be reflected in human mind relative to social and natural processes.

The greatest danger to the discovery of the efficient path of knowing of ontological elements and the understanding of the applications that may be required of them is ideological and scientific credulity that imprisons creative imagination in the walls of familiarity of thought and reasoning in the comfort zone of intellectual ignorance and arrogance. This ideological and scientific credulity finds an expression in the classical paradigm, with the principle of exactness of information, information representation and claims to absolute truths, on the basis of methodology of exact rigid determinations that cement the foundations of current scientific knowing in the knowledge-production process. The danger may be diminished, however, by developing cognitive habits of different forms of mendacity in order to create a more robust foundation of fuzzy paradigm composed of its laws of thought, logic and mathematics with the principles of inexactness of information, symbolism and claims of conditional truths on the basis of methodology of inexact flexible determinations of knowledge. In this respect, and in the development of the fuzzy paradigm, enhancement of its logic and expansion of its mathematical domain, we are thankful of all the researchers and scholars who have freed themselves from the ideological grip of the classical paradigm with Aristotelian principle in order to work on the frontier of fuzzy phenomena in all areas of the knowledge enterprise. This monograph has benefited from their contributions as well as those working in the classical paradigm as seen from the references.

I express my thanks to my admirers and critics whose positive and negative reflections have reinforced my theoretical and philosophical convictions and made this work enjoyable to the end. My thanks also go to all my graduate students in my courses in economic theory, mathematical economics, operations research, cost-benefit analysis and international economics for their tolerance to the idea that the use of the classical system of thought runs into important epistemic limitations when and where qualitative dispositions are to be dealt with and when the system is as complex as socioeconomic one. Similar thanks go to Professor Momoh and his students in the electrical engineering who were working on the problems of power system, as well as fuzzy controllers, for allowing me to share my works on the fuzzy paradigm. This monograph has also benefited from my seminar works on the current methodological frontiers of the fuzzy paradigm with some students in the Computer Science Department and African Studies Department at Howard University. Special thanks go to Professor Rey Madoo for his continual encouragement and motivation. This monograph has also benefited from the works of Drs. Kwame Nkrumah, Cheik Anta Diop, and W.E. Abraham by way of motivation. All errors rest with me.

# Preface

*There is nothing more dangerous for new truth than old delusion*
W. Goethe

*It is a fact that all general theories grow from studies of particular problems and they do not have any meaning unless they can explain more specific questions and bring some order in them.*
R. Courant

*Furthermore, any genuine science must, even as science be relevant to action. But natural science itself, in its relevance to action to which it is relevant is not merely action upon natural objects, which are inert and passive, but is also action by a subject who does act and who consequently has a fundamental nature categorically opposed to that of the object matter of action. Social science must also but relevant to action and to the kind of action to which it is relevant. But this is not action upon an inert, passive object matter* [R16.17, p.132].

*Today nobody any longer denies that on account of the insufficiency of our senses measurement is never precise in the full sense of these terms. It is only more or less approximate. Besides, the Heisenberg principle shows that there are relations that man cannot measure at all. There is no such thing as quantitative exactitude in our description of natural phenomenon. However, the approximations that measurement of physical and chemical objects can provide are by and large sufficient for practical purposes. The orbit of technology is an orbit of approximate quantitative definiteness* [R18.14, p. 62].

## I. The Monograph

At the level of meta-theoretic investigation on the progress of knowledge development, the task of meta-theoretic analyses is to search for a general principle of the knowledge production irrespective of whether the knowledge production is about either nature or society or both. All systems of knowledge production must be seen as human attempts to strip itself from ignorance and uncertainty by providing a rational account, not because we have ignorance and uncertainty, which we do, but to answer the question: what justification do we have to claim knowledge items and certainty of them for understanding and practice? In this connection, any knowledge system demands a set of rules, a paradigm so to speak, which forms the information processing machine as well as ideological boundaries to protect it and provide the belief system to claim their validities by cognitive agents.

In this monograph an argument is offered that the uncertainties about knowledge production, discovery and growth are rooted in a critical and better understanding of knowledge of uncertainties and their relationship to information, methods of cognition and the effective utilization of these methods. The limited knowledge of these uncertainties is due to the failure of knowledge agents to understand the relationship among information, knowledge and ignorance on their part in linking the epistemological space, which is under the control of decision-choice agents in the acts of knowing, to the ontological space that exists outside the control of cognitive agents. At the level of knowing, the knowledge about uncertainties must involve questions about defective information structure which is composed of information vagueness that gives rise to fuzzy uncertainty, possibilistic belief system and fuzzy risk on one hand; and then information limitation that gives rise to stochastic uncertainty, probabilistic belief system and stochastic risk on the other. All of these involve complex interactions of ontological and epistemic elements that involve the processing of defective information structure.

At the level of cognition, P. N. Fedoseyev reflecting on Scientific Cognition Today, its Specific Features and Problems has this to say:

> The development of natural sciences and philosophy has shown that it is impossible to explain the whole wealth of the relationship between theory and experience and answer many important questions posed by the progress of science (particularly the question of character of scientific revolutions) if one confines oneself to formal-logical analysis in considering these problems. A far more comprehensive approach embracing all the essential elements in man's search for knowledge is required [R8.21b, p.15].

The general intent of the monograph is to show the origins of paradigms relative to the assumed information structure and make a case for the use of fuzzy paradigm and its laws of thought in relation to the classical paradigm and its laws of thought and the applicable areas of knowing.

## II. Experience in Cognition

My intellectual path and growth have taught me new things as well as forced me to unlearn some methods of reasoning in logic and mathematics and search for other new methods of theoretical understanding. In economics, I learned the dual concepts of supply and demand, cost and benefit, input and output, substitution and transformations, quality and quantity, negative and positive, open and closed and many others. All these reflect the organic principle of the principle of opposites. Here, I became convinced that events of life present themselves in dualities with continuum for relative understanding. However, it is sometimes difficult if not impossible to simultaneously think in conceptual duality with a continuum of the same phenomenon where the new and the old are simultaneously present in the sense that the seed of the new is always in the womb of the old and the reminisce of the old resides in the growth of the new; and where every accepted truth has some elements of falsehood and every falsehood, in turn, has elements of truth. The conditions of opposites, duality, polarity and continuum seem to constitute the basic principle of ontological and epistemological transformations. Respectfully, the elements of the disappearance constitute either the costs or the benefit; and the elements of the emerging new constitute either the benefits or the cost of the continual substitution-transformation processes in nature, society and knowledge production. It seems that the understanding of natural and social processes can be undertaken under the general principles of opposites where each phenomenon contains its opposites in relative proportions to give its identity.

   In the development of the meta-theory of knowledge, a counsel was taken from a number of African traditional advices. No one is without knowledge except the one who asks no questions; Knowledge is like a garden: if it is not cultivated, it cannot be harvested; not to know is ignorance and not to want to know is mental slavery. The wisdom of today is a folly of tomorrow where the breaking day has wisdom and the falling day has a simple experience; if you fear something you endow it with power over you because too much fear does not create freedom but slavery and bondage. The adherence to these advices unlocks the creative essence of intellectual labor for continual process of the step-by-step progression to overcome the ideological boundaries that restrict human progress in knowing. This creative essence of intellectual labor has an active categorial moment of information-knowledge transformation to establish powerful flames of ideas to shine the dark sports of the knowledge house that death is incapable of extinguishing.

## III. Decision-Choice Space and Knowledge Production

Our knowledge-production process to reduce human ignorance about nature and society is an outcome of decision-choice modules working with information as input and paradigms of thought as processing tools to understand the complexity of the principles of opposites, as revealed in human cognition about an epistemic element. Information relates to the potential reduction of ignorance through the production of knowledge with information as an input and decision-choice modules as a technical processor. The precision of the claimed knowledge and sureness attached to it is a reflection of the attachment of same properties to information and the paradigm of thought and correctness of its use. The process is an input-output phenomenon that generates principles through which knowledge is claimed to reduce ignorance. In this respect, we may speak of the knowledge content of information just as we speak of the information content of an event. The latter is directed toward the question of how much information is available from an event or a phenomenon and the former is directed to the question of how much knowledge is contained in the information of the event or the phenomenon. Both of them are cognitive activities.

The first question is answered by acquaintance and the construction of *sense data* that form the primary category for initializing the epistemic journey. The second question is answered with a paradigm of thought in deriving the knowledge content as a derived category that must be related to the answer to the first question. The relationship between the acquaintance and information on one hand, and paradigm of thought and knowledge on the other hand is the subject matter of the theory of knowledge square. The theory of the knowledge square, thus, presents a meta-theory on the general theory of knowledge systems in relation to its development and growth by examining the path of knowledge-production activities in general. Its task is to examine the path of epistemic activities that allow us to claim knowledge content to establish an isomorphic relation between an epistemic element and an ontological element. The path of knowledge search is the same in that the knowing activities take place in the same epistemological space that defines the highway to the universal knowledge house. What then separates knowledge areas into science, non-science, exact science and inexact science? The answer to this question is an element in the set of objectives of the theory of the knowledge square.

Through the examination, conditions are derived to establish the fuzzy rational foundations for demarcation of knowledge areas, in particular, science and non-science on one hand and exact science and inexact science on the other hand. In the general education, a belief is established regarding the existence of exact and inexact sciences where our number system is claimed to be exact. Similarly our symbolic system is also claimed to be exact. This claim of exactness in science, classical mathematics and classical logic does not have indisputability. The indisputability rests on the set of assumptions imposed on the epistemological space whose information characteristics contain qualitative and quantitative dispositions. The nature of the set of assumptions depends on the organization of knowledge production. The fundamental structure of the epistemological space is abstracted to establish the framework of the debate between the classical logic and its mathematics and the intuitionist logic and it mathematics leading to the fuzzy rational foundation of exact and inexact sciences.

## IV.  The Organization of the Monograph

To accomplish the task at hand and the set of objectives the monograph is organized in eight chapters. Chapter one is used to present the central problem of interest to the development of the theory of the knowledge square. It discusses the epistemic point of entry and point of exit into exact and inexact sciences and the need for a meta-theory of knowledge. First, an explicit distinction is made between the ontological and epistemological spaces. The distinction allows us to define and specify the ontological elements that are distinguished from the epistemic elements. The condition of separation is specified in terms of information structure where the ontological space is defined by non-defective (perfect) information structure as an input into the activities of non-cognitive agents. The epistemological space is defined by defective information structure as an input into the activities of cognitive agents. The definition of the concept of defective information structure is linguistically and symbolically offered.

The rest of the chapter initializes the building blocks of the theory of the knowledge square by introducing the concept of the potential space in relation to the meta-theory on the theory of knowledge where the defining properties of the potential space are established. The ontological elements and the ontological space are taken as existing in a perfect state and hence there are no imperfections and there are no vague ontological elements. The epistemic elements and the epistemological space are under the information-knowledge processing capacity of the cognitive agents. Vague objects are the characteristics of the epistemological space that that amplifies human ignorance and uncertainty.

Chapter 2 is used to introduce the concept of the possibility space as the second analytical space of information processing activities of cognitive agents. The possibility space is introduced as composing of elements that are possible candidates for knowing. The definitions of the concepts of possibility and possibility space are provided in terms of information structures. The possibility space constitutes an initial conceptual and analytic bridge between ontology and epistemology at the level of information-knowledge phenomenon. The elements in the possibility space are called *possibilistic epistemic elements*. They are established through information generation and the principle of acquaintance that involves the non-enhanced and enhanced senses. The information connecting the knowledge-potential space and the possibility space is incomplete and vague. The information structure with the characteristics of incompleteness and vagueness is termed *defective information structure* that becomes an input into the decision-choice system of the knowledge-production process. The possibility space, on the other hand, is constructed by a process involving the use of membership characteristic functions to define conditions of belonging to categories that may justify conditions of possible knowledge. The component of the vagueness of the defective information generates possibilistic uncertainty of conditions of possibilistic epistemic elements where such uncertainty is argued to characterize all knowledge areas. The presence of vagueness defines degrees of belonging whose analytical structure is argued to require the need and use of fuzzy paradigm.

The concepts of possibility, possibility space and possibilistic epistemic elements are defined. The possibility space is argued to carry with it the condition of inexactness and hence contains inexact epistemic elements. The defective information as the essential attribute on the possibilistic epistemic elements is represented by fuzzy symbolism. The discussions link epistemology and ontology to the nature of analog and digital spaces of acquaintance and language and then to the role of decision-choice actions in knowledge production. The problem representation of the defective information structure leads to the discussion on the nature of classical mathematical and logical variables in relation to linguistic variables and how they interact with fuzzy variables. The relationships among the possibility space, the possible world space and the knowledge-production process are discussed and connected to the classical and fuzzy paradigms. A discussion is then made on the role of mathematics in classification of science and the epistemic process.

The chapter is concluded with the discussions on the possibility space, the formation of the primary category and the derived category in the knowledge-production process. The distinction between explanatory and prescriptive sciences is made where their roles in the knowledge-production process are defined and explicated in terms of decision-choice rationality in the knowledge-production process. The possibility space is used to initialize the epistemic algorithm while the defective information structure and the nature of cognitive processing define the epistemological terrain. The introduction of the defective information structure strips off the a-prior condition of exactness in knowledge and science. In this way, a framework is created to provide the epistemic conditions for the use of the fuzzy paradigm that may or may not support the conclusions from the classical paradigm.

Given the possibility and the possibility space, Chapter 3 introduces the concepts of probability and probability space from which their roles to the theory of the knowledge square are made explicit. The nature and the need for probabilistic reasoning in the process of knowing are developed and shown to be related to all areas of exact and inexact sciences and other knowledge areas. The concept of probability is explicated with discussions on the conditions of its measurement and related to the concepts of possibility and possibility space in order to specify the *probabilistic epistemic elements*. From the possibility space the probabilistic epistemic elements are constrained by possibility and impossibility as viewed in terms of duality under the principle of fuzzy continuum. The probability space that is conceptually presented here incorporates exact and inexact probability values and related to *fuzzy-random process*. In this respect the probability space in its general form is rightly called the *fuzzy probability space* where the probabilistic epistemic elements are called *fuzzy probabilistic epistemic elements*.

The concept of probability is related to the concept of information incompleteness which defines the other component of the defective information structure on top of information vagueness which the other component of defective information structure discussed in the possibility space. The effects of the defective information structure in defining and specifying the epistemological space are made explicit and then related to the paradigm developments. The conditions of paradigm development are related to the classical paradigm with its laws of thought, logic and mathematics. Similarly, the conditions of paradigm development are related to the fuzzy paradigm with its laws of thought, logic and mathematics. These conditions are connected to the concept of exactness-inexactness duality and dualism of epistemic actions over the epistemological space. The conditions of exactness-inexactness dualism are linked to the principle of the excluded middle of the classical paradigm which works with *exact epistemological space* defined by an exact and complete or incomplete information structure. The conditions of exactness-inexactness duality are linked to the principle of the continuum of the fuzzy paradigm which works on an *inexact epistemological space* defined by an inexact and complete or incomplete information structure. The exact classical probability space is abstracted from the exact epistemological space; the fuzzy probability space is abstracted from the inexact epistemological space.

The probability space is then linked to explanatory and prescriptive sciences in terms of *what there is* and *what ought to be* in order to specify the epistemic cord that connects to the space of the epistemic reality with defined *fuzzy risk* and *stochastic risk*. The fuzzy and stochastic risks are connected to the concept of an irreducible core of general uncertainty and used as fuzzy-stochastic conditionality for knowledge and truth verification, falsification and corroboration in order to construct the space of the actual. The uses of the fuzzy and stochastic risks extend to scientific, technological, engineering and mathematical systems.

Chapter 4 is used to present the concept and the space of the epistemic actual as the fourth building block in the theory of the knowledge square. The conditions and the properties of the space of the epistemic actual and its relationship to reality are offered and distinguished from the ontological actual. The manner in which the probabilistic epistemic elements enter into the space of the epistemic reality is discussed under the presence of ambiguities and inexactness. The analytical process of conversional linkages requires the explications of the key concepts of epistemology and methodology where the similarities and differences are made explicit and connected to the concepts of cognitive frame, toolbox frame and epistemic frame as defining the space of methodology.

The thinking instruments of analytic and synthetic logics are structured for understanding in the applications of fuzzy and classical paradigms. The differences and similarities in the paradigms are made explicit through the concepts of discreteness and continuum as they relate to exact and inexact symbolism in information representations and logical manipulations under defective information structure. The framework is analytically connected to the relational structure of nominalism, constructionism and reductionism in the knowledge-development processes. The meta-theoretic frame leads to a situation where a distinction is made between ontological information and epistemological information where the game of the knowledge-development is to show the identity of the two information sets of the same epistemic element. The central analytical positions of the theory of the knowledge square as a meta-theory on the theory of knowledge are summarized for reflections leading to the discussions on the space of the paradigms of thought and emergence of the fuzzy and classical epistemic systems.

In chapter 5, the analytical frame for the development of paradigms of thought over the epistemological space is presented. The concept of a paradigm is introduced, defined and explicated in relation to the defective information structure that allows conceptual constructs of exact and inexact epistemological spaces. The exact epistemological space is related to the conditions required for the development of the structure of exact symbolism and the logical rigidity of the classical analytics and their relationship to the classical paradigm, its laws of thought and mathematics of reasoning. Similarly, the inexact epistemological space is related to the conditions required for the development of the structure of inexact symbolism and the logical flexibility of the fuzzy analytics and their relationship to the fuzzy paradigm, its laws of thought and mathematics of reasoning. The conditions determining the foundations of both the classical and fuzzy epistemological spaces relative to information representations and knowledge developments are presented with some analytical clarity of their differences and similarities.

The foundations of the classical paradigm are connected to dualism, and those of the fuzzy paradigm are connected to duality and related to the fundamental continuum where the relationships are then discussed in terms of negative and positive characteristic sets that are used to establish the conceptual identities of the epistemic elements relative to the ontological elements. The natures of the degrees of exactness and inexactness embodied in both classical and fuzzy laws of thought are brought forward and related to the classical epistemic rationality and fuzzy epistemic rationality. The role and effects on quantity-quality-time space are examined where the structures of the paradigm developments are related to the concepts of fuzzy and stochastic conditionalities. The chapter is concluded with the summaries of the epistemic pillars of classical and fuzzy paradigms.

Chapter 6 is devoted to the critical examination of the relationships among information, fuzziness, and science, the theory of the knowledge square and the claims of exact sciences. Here, the space of knowing is partitioned into science and non-science and their concepts explicated. The space of science is partitioned into exact and inexact and the conditions that define their boundaries over the general epistemological space are examined. The chapter is also used to discuss the foundational conditions of exact and inexact symbolisms and then related to definitions, explications and the claim of the existence of exact science. The concept of irreducible core of inexactness is introduced through the fuzzy process where definitional and explicator sets are used to link common language to the language of science.

The axioms of definability and explicability are presented and explained in relation to exact and inexact symbolisms for information representation through the introduction of definitional and explicator functions. The conditions of the existence of an exact symbolism in science, logic and mathematics are presented and linked to a fixed point theorem. Definitional and explicator fixed point theorems for the existence of exact symbolism are presented where the theorem of non-existence of exact symbolism is advanced and proven. The theorem of non-existence of exact symbolism presents a challenge to the classical paradigm as a general information processing instrument for thought and knowledge production.

This challenge leads to the discussions of the affirmation of inexact symbolism and the emergence of intuitionist paradigm with its logic and mathematics and then to the fuzzy paradigm and its logic and mathematics.

Chapter 7 is utilized to examine the relational conditions of the defective information structures for the cognitive works in the epistemological space and how these works connect to the ontological space through the understanding of the concepts of matter, energy and information as a fundamental relation that connects the ontological space to the epistemological space through the conversional power of the information process. The characteristic sets that generate inputs into the information process in the ontological space is taken to be exact while the characteristic sets that generate inputs into the information process in the epistemological space is taken to be inexact by the nature of human limitations. There are discussions for where the epistemic process works for the conditions of exactness to the conditions of inexactness. The nature of defective information structure over the epistemological space is argued to hold for all knowledge sectors including science and hence any segment of science cannot be claimed to be unquestionably exact.

It is the presence of defective information structure, as characterizing the epistemic space, that provides the driving force for the development of logics and mathematics of possibility and probability to obtain the conditions and measurements of fuzzy and stochastic conditionalities and the associated risks in knowledge systems and applications leading to the introduction of the irreducible core of epistemic risk. The defining characteristics of science, non-science, and exact science are discussed and related to the nature of information structure assumed, the objectivity claimed and quantitative disposition imposed. The chapter is concluded with the discussions on the relationships among knowledge items, conditions of justification, corroboration and verification in the acceptance processes for an epistemic item to enter the epistemic reality.

Chapter 8 is the concluding chapter for the development of the monograph which is devoted to a meta-theory on knowledge and the examination of conditions for the development of the fuzzy rational foundations of exact and inexact sciences. The identities of knowledge and science are examined through the introduction of empirical and axiomatic conditions in defining and establishing the primary category for the foundation and the initial information for categorial conversions. The relative concepts of empirical ontology, empirical epistemology, axiomatic ontology and axiomatic epistemology are introduced in relation to the initial information structure accepted to allow the comparative examination of *what there is* and *what the knower knows*. These concepts lead us to examining the objectives and classification of science into exact and inexact sciences and then into explanatory and prescriptive sciences. The discussion leads to a summary case in support of the need for fuzzy paradigm as a generalized information processing machine over the epistemological space in the knowledge production. Scientific and non-scientific methodologies are discussed in relation to toolboxes available for information-knowledge transformations. The manner in which cognitive agents are integrated into the epistemic process is discussed.

# Prologue

Decision-choice agents function as cognitive agents before undertaking decision-choice action in all spaces of human endeavor. The work of cognitive agents is an information collection and processing into a knowledge system. In other words, there is a cognitive mapping from the information space into the knowledge space where such a mapping is an input transformation module. The work of the decision-choice agent is the use of the obtained knowledge as input into decision-choice action. The idea of knowledge production as an output from information processing suggests to us that information is not knowledge; it has to be refined to take away noises that create unsureness in order to improve dependability of use in the general decision-choice process. This unsureness comes under the general cover of uncertainty. There are two important questions that must be answered in relation to total uncertainty. Where does this unsureness enter the space of operations of cognitive agents? What is the meaning of uncertainty and what is its definition? Can the concept of uncertainty be explicated to give it a scientific content that covers the essential total aspects of human ignorance in knowledge and decision processes?

## I. The Concept of Uncertainty in General

Unsureness is associated with doubt of desired outcome of processes in nature and society. Doubt is a cognitive state of being that gives rise to the feelings (thoughts) of many possibilities and probabilities of the realization of a process. The feelings (thoughts) of many possibilities and probabilities, without a clear knowledge of what possible and probable outcome would be realized, give rise to uncertainties.

### Definition: Uncertainty

Uncertainty is a cognitive state of being whose information and processing capacity constrain the abilities of cognitive agents to derive exact knowledge about possibilities and probabilities of realizations of outcomes of processes in nature and society, as such, every claim of knowledge item resides in the true-false duality with partial state of truth with a degree of false support that creates partial confidence in the decision agent's choice actions.

The understanding of uncertainty in language and science has claimed the efforts of many important minds and yet its conceptual definition and explication seem to elude clarity in the knowing and decision-making processes that must be related to knowledge which also contains cognitive awareness. At the level of awareness, uncertainty is an information phenomenon. At the level of knowing, uncertainty is phenomena of defective information structure, and paradigms as information-processing modules. At the level of choice, uncertainty is a decision-choice phenomenon. The uncertainties about knowledge production may be attributed to the deficiency in the understanding of the nature of the epistemological space that carries the information input into the epistemic processing system where actions and outcomes can reduce the ignorance of the cognitive agents. In this framework, information is a primary category of the epistemic process; knowledge is its derivative, and ignorance is a derivative from knowledge which gives rise to uncertainty. The general framework of human action is such that uncertainty is related to ignorance which is related to information as an input for knowledge production. The outcomes of knowledge production are the results of activities in the decision-choice space. Uncertainty, in relation to ignorance of cognitive agents, is related to a number of conceptions of types of ignorance rooted in the nature of the information structure that is believed to be available as input into the knowledge-production process.

In this respect, the first encounter of uncertainty may be seen at the primary level of knowing in terms of information characterization as input into the epistemic process over the epistemological space. It is useful to see uncertainties as the disparities between ontological information and epistemic information regarding any particular phenomenon at the threshold of knowledge production. In this way, uncertainties are seen as characteristics of the information-knowledge activities of cognitive agents in the epistemological space relative to the ontological space. The differences are the resulting reflections of the nature of the epistemological space, and due to two factors of *incomplete reception* of the ontological information characteristics (quantity of information) on one hand, and vagueness and ambiguities of the received information signals (quality of information) as they are cognitively coded and linguistically represented for the epistemic processing machine (the paradigm of thought).

It may be pointed out, as it has been argued elsewhere; there is neither ontological uncertainty nor ontological risk [R11.22]. What we have are epistemological uncertainty and risk. What are present in the ontological space are continual substitution-transformation processes with a continuum in the quality-quantity duality and neutrality of time as the environments of the ontological elements change to define new conditions of relationality. In human actions, however, we are presented with epistemological uncertainty and epistemological risk that flow from ignorance created from the defective information structure for epistemic processing by cognitive agents. The problem, in general, is to understand the epistemological uncertainty, how the uncertainty is transformed into risk in human action and to design solutions to reduce the uncertainty and the corresponding risk. Over the epistemological space, therefore, the less defective is the information structure, and the more efficient is the paradigm of thought, the less is the cognitive ignorance and the greater is the epistemic certainty. From the viewpoint of logic as a guide of reasoning and thinking, Bertrand Russell's advice is useful.

> *The philosophical importance of logic may be illustrated by the fact that this confusion--- which is still committed by most writers---- obscure not only the whole study of the forms of judgment and inference, but also the relations of things to their qualities, of-concrete existence to abstract concepts, and of the world of sense to the world of Platonic ideas. Peano and Frege, who pointed out the error, did so for technical reasons, and applied their logic mainly to technical developments; but the philosophical importance of the advance which they made is impossible to exaggerate.*

The importance of the concept of quality in the Russell's reflection cannot be overlook. It is an important component of the fuzzy logical and analytical system since it introduces subjectivity and judgment into the knowledge production system by internalizing the cognitive agent. It is also, here, that fuzzy mathematics and fuzzy logic are inseparable in the fuzzy paradigm of thought.

## II. Uncertainty and Qualitative-Quantitative Dispositions

The information received through the acquaintances of the characteristic set has qualitative disposition and quantitative disposition, both of which must be combined into one in the knowledge-production process to compute the knowledge content that will support decision-choice actions by decision-choice agents. In this respect, we separate knowledge-production activities from non-knowledge production activities of decision-choice agents. Decision-making in the field of knowledge-production entails uncertainty. Similarly, decision-making in the use of the results of knowledge production entails uncertainty viewed in terms of hesitation of action due to lacks of surety of outcomes.

The qualitative disposition of information is associated with vagueness, ambiguities, imprecision and others under the general structure of fuzziness over the epistemological space. The information containing these characteristics is termed fuzzy information structure, which gives rise to fuzzy uncertainty and fuzzy risk. The presence of fuzzy information structure is the result of acquaintance, human limitations in observation, human langue in classification and communication and others in developing the sense data for processing. In a broad general way, therefore, qualitative disposition of information may be associated with the source reception of information where real situations, as appear to cognitive agents, are not crisp and completely deterministic. It originates from the construct of knowledge possibilities. This is so irrespective of whether one is dealing with systems of simplicity or complexity. The difference in terms of crispness and rigidity is in degrees and varies over subjects and phenomena.

The quantitative disposition of information is associated with incompleteness of the volume of information in terms of cognitive agents receiving full information characteristics about any phenomenon. The quantity of information is a volume phenomenon of acquaintance, which may be related to some form of measurement with a specific unit. It originates from the construct of knowledge probability that may be available for practice and action. At the level of information collection, therefore, there are two sources of uncertainties. One type of uncertainty is due to qualitative disposition of information termed *fuzzy uncertainty* and corresponding to it is the fuzzy risk. The other type of uncertainty is due to quantitative disposition of information termed as *stochastic uncertainty* and corresponding to it is the stochastic risk. The fuzzy uncertainty flows from possibilistic information while the stochastic uncertainty flows from probabilistic (incomplete) information. The analytical task is how to identify and combine then for understanding human decision-choice actions and the risk of practice.

At the level of knowledge-production, and given the initial information conditions, there is an uncertainty that is associated with the selected paradigm of the knowledge production. This uncertainty will be characterized as *epistemic uncertainty* that is associated with the paradigm used in the information processing to obtain thought or knowledge. Here, we have *fuzzy epistemic uncertainty* due to the use of fuzzy paradigm, its laws of thought and mathematics. There is the *classical epistemic uncertainty* due to the use of the classical paradigm, its laws of thought and mathematics. At the level of decision choice actions, there is an uncertainty due to the uses and misuses of the knowledge as an input into the decision-choice action that may require judgment as decision-choice agents pass through the penumbral regions of decision-knowledge interactive domains. This is the decision-choice uncertainty. All these types of uncertainty are defined in the epistemological space, and hence from the viewpoint of the meta-theoretic construct in the current work. There is no ontological uncertainty or risk. The types and the paths of uncertainty are shown in Figure A.

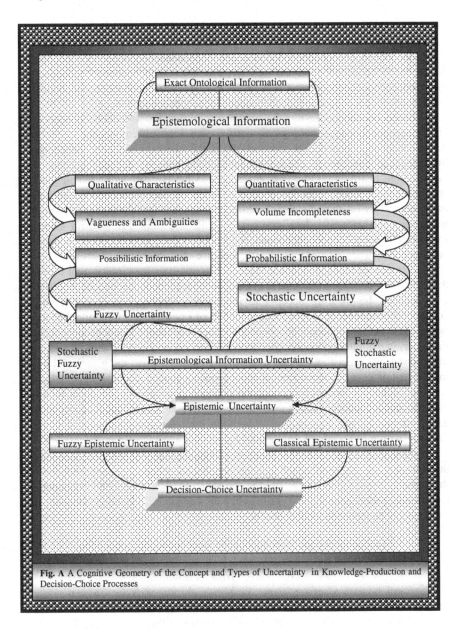

**Fig. A** A Cognitive Geometry of the Concept and Types of Uncertainty in Knowledge-Production and Decision-Choice Processes

Any demarcation of science must take account of the nature of information structure that is assumed, the information representation needed, and the paradigm required for processing the information to attain the knowledge. The process must also account for the nature of the corresponding uncertainty and the corresponding epistemic risk, and how the uncertainty and risk differ in demarcations relative to the removal of the lack of surety about the knowledge. These will indicate to cognitive agents the type of the epistemological space that is assumed for the information processing. It must be understood that the epistemological space is a decision-choice created that includes information structure and the cognitive agents as part of knowledge dynamics. The cognitive and decision-choice agents do not exist outside the epistemological space which is their creation. Neither do they exist outside the epistemic process for knowledge development. The nature of the epistemological space is transformed by cognitive agents, who are in turn transformed by their activities in search of knowledge for effective decision-choice actions. The epistemological controversies, their progresses and resolutions in the knowledge production, can only be resolved in the decision-choice space that admits of subjectivity, objectivity, quality and quantity in order to construct a falsification, justification and corroboration of what constitutes knowledge in the epistemological space. This knowledge is never the final one, and is constantly being refined by fuzzy and stochastic tunings as the epistemological information is improved. The cognitive decision-choice agents constitute the primary category from which the epistemological space is derived by their actions and the epistemic process is constructed from their decision-choice modulus.

Over the epistemological space, therefore, a revolutionary intellectual program that satisfies a new methodological system for operations is not merely negative to a system of the existing acceptable intellectual ideology. It is not a simple conceptual refutation of a dying social system of knowledge production, but it is a positively creative epistemic theory, derived in establishing a new guiding light to the emerging intellectual order that will include the vestiges of the old order. In this respect, we must see the unfolding of the positive enveloping path of human intellectual progress as the forces of ideological limitation of old intellectual order crumble under its own weight, with the emergence of a new paradigm to assert itself in the epistemological space, to finally liberate great minds from mental habits of intellectual slavery that effectively imprison imagination and creative thinking within the walls of intellectual habits and ideological constraints without a breathing space. The continual fuzzy-stochastic tuning and retuning demand the development of an intellectual culture of doubt, where we link the past to the future through the eyes of the present, for creative improvements by examining paradigmatic and philosophical categories. In this respect, the reflection of Bertrand Russell is useful

> And since the philosophies of the past belong to one or other of few great types--- types which in our day are perpetually recurring--- we may learn, from examining the greatest representative of any type, what are the grounds for such a philosophy. We may even learn, by observing the contradictions and inconsistencies from which no system hitherto propounded is free, what are the fundamental objections to the type in question, and how these objections are to be avoided. But in such inquiries the philosopher is no longer explained psychologically; he is examined as the advocate of what he holds to be a body of philosophic truth.

This Russell's reflection is amplified by Kwame Nkrumah's reflection that simply states:

> *The critical studies of the philosophies of the past should lead to the study of modern theories, for these latter, born of the fire of contemporary struggles are militant and alive....Philosophy, in understanding human society, calls for an analysis of facts and events, and an attempt to see how they fit into human life, and so how they make up human experience. In this way, philosophy, like history, can come to enrich, indeed to define the experience of man* [R14.85, p. 3]

A search for understanding the rise of different paradigms of thought over the epistemological space has been the driving force of this monograph. The main concern, at the moment, is the understanding of the relative positions and strengths of classical and fuzzy paradigms, and the kinds of thinking systems that they engender. The increasing expansion of research in fuzzy system, logic, mathematics and technology, and the need to solve the problems of vagueness, quality and inexact symbolism, require a philosophical action and explanation of their scientific directions and specifications of the applicable areas of their applications in all areas of knowledge production. This, we hope, will allow an attempt to provide a *unified epistemic frame* for further research on inexact epistemological space, and the development of the fuzzy paradigm, its logic and mathematics, and their uses in the knowledge production, especially in the areas where quality and vagueness are inseparably connected to the phenomena, and where quantitative transformations of quality are not available. The hope is that, a path will be opened to integrate knowledge sectors through an epistemic method of unified information processing capability where the knowing is integrated as part of human behavior. A reflection on the problem of knowledge production as an automatic self-control system is presented by the following statement by E.C. Harwood.

> *In their attempt to know more about 'knowledge,' Dewey and Bentley have regarded knowing as an aspect of human behavior. No inner knower, or mind, or soul that does the knowing is assumed. They take man as they find him behaving in his cosmos or universe, never apart from it. Likewise, they take whatever is known in the cosmos or universe as they find it, never in isolation from it* [R8.26, p.8].

In the development of the meta-theory of knowledge, a counsel is taken from a number of African traditional thoughts. No one is without knowledge except he who asks no questions; Knowledge is like a garden: if it is not cultivated, it cannot be harvested; not to know is ignorance and not to want to know is mental slavery. The wisdom of today is a folly of tomorrow where the breaking day has wisdom and the falling day has simple experience. This counsel fits into the notion that every self-contained paradigm must be an epistemic program. The epistemic program must be constructed from an alternative representation of the information system with the rules of combination. The information representation and the rules of combination must be positive and organic in such a way as to replace or absolve existing paradigms as well as to be contestant to possibly emerging paradigms. The first, necessary and important condition for the paradigm's development is its acceptance. Its development must be confined to those that understand the existing paradigms and their analytical difficulties and how the new paradigm can resolve the analytical difficulties and existing paradoxes. It is under this framework that the theory of the knowledge square is developed in this monograph. It presents aspects of *computational epistemology* as a guide to the understanding of information-knowledge structure in the processes of knowing. As used here, the primary purpose of the computational epistemology is to provide a qualitative and quantitative framework of computable logical system to justify epistemic claims, establish knowledge distance between an ontological element and epistemic element of the same phenomenon, and abstract an epistemic conditionality in the support of epistemic claim. Such a framework that deals with the problem of quality and quantity finds its analytical tools from the fuzzy paradigm, its logic and mathematics. It belongs to the thought systems of computational philosophy which will help us to understand the complex system of knowledge production, reduction of ignorance, uncertainty and risk in human actions in all areas of human endeavors.

## III.  A Note on Epistemic Conditionality and Knowing

The emphasis on the epistemic activities in the knowledge production system must center around the understanding of the factors that affect the thinking system as a self-correcting process which enlarges the mind in the cognition but not simply on the orthodoxy or heterodoxy. Scientific philosophy must assist us in search of efficient path fot the knowledge-production process. The search for the optimal path of learning-thinking process must lead to the development of the needed logical framework for the establishment of the initial conditions, paradigm of thought, and *epistemic transversality conditions* for the optimal path of the cognitive process. The epistemic transversality conditions are simply the terminal conditions that help to distinguish a knowledge item from other admissible knowledge claims at the end of the temporally epistemic journey, which is induced by a paradigm, given the initial conditions. There are two types of epistemic transversality conditions that are of special interest to us. They are *classical epistemic transversality* conditions and *the fuzzy epistemic transversality* conditions. The classical epistemic transversality conditions require the epistemic risk to converge to zero through the stochastic tuning process where the initial conditions of limited information exactness and the classical epistemic transversality conditions are taken as sufficient to identify and claim a knowledge item with *stochastic conditionality* that tends to zero, as the quantitative information space widens within the classical paradigm with its logic. The fuzzy epistemic transversality conditions require the epistemic risk to converge to a positive value of the core of the risk through the fuzzy-stochastic tuning process where the initial conditions of limited information inexactness and the fuzzy epistemic transversality conditions are necessary to identify a knowledge item with fuzzy-stochastic conditionality, as the quantitative information space widens and the space of vagueness of information decreases within the fuzzy paradigm with its logic and mathematics.

The initial conditions, the paradigm and the epistemic transversality conditions may be related conditions of simplicity and complexity on one hand and relevance and irrelevance on the other hand. It is somehow accepted that one of the principal considerations guiding the process of creating theoretical frames is simplicity. Behind this simplicity is a search of elegance. Behind this search of elegance, are complex inexact information inputs that are many times difficult to represent as simple and exact, except by imposing a set of assumptions. The set of assumptions may render the resulting theory simple and irrelevant, especially when there are simultaneous existing of quality and quantity with continual transformations as time proceeds. Simplicity involves conception of phenomena (such as particles, points and instantaneous changes) with fixed quality. Complexity involves conception of phenomena (such as living organism, social organism, and gradual transformations) with both quantitative and qualitative changes. The main task of this monograph is the analysis of the analytical structures of complex and simple theories, as they relate to initial information conditions, paradigms, and transversality conditions, for identifying epistemic truths and reality in all areas of knowledge production.

We have kept in mind that the effective use of a paradigm requires an evaluation of the environment of the theoretical and applied works of one's subject area of study and how this area is related to other areas of the knowledge production. It, further, requires a critical analysis of the social circumstances surrounding the particular area of cognition, and its relative contributions to the general system of knowing. All these constitute parts of the information set, and indeed, they constitute important information characteristics for the initialization of the epistemic journey over the epistemological space. It is the subject areas of knowing that define the variable initial starting points as well as the varying terminal points of the epistemic journey.

# Contents

# Chapter 1
# The Theory of the Knowledge Square and the Information Structure: The Points of Entry and Departure

In a number of my writings on philosophy of science, the theory of knowledge and foundations of fuzzy laws of thought, I view them from the vantage point of the theories of economic behavior, complexity of social dynamics, self-correcting systems, decision-choice systems and synergetic behavior of systems rather than as a philosopher or a mathematician. From this vantage point, I introduced an analytical concept of the *knowledge square* to allow me to deal with the problems of the relative positions of the critical concepts of *exactness and inexactness* in the general knowledge-production process. The problems of exactness and inexactness as seen in the exact-inexact duality are related to quantity and quality of information structures in categorial conversions and epistemic transformations of information signals into knowledge elements in the epistemological space [R2.3] [R2.10] [R11.21] [R11.22].

The discussions in this monograph are to make explicit the analytical strengths of the theory of the knowledge square and its conceptual path for knowing. It will also indicate the required logical structure of knowing. The theory of the knowledge square is to assist in assembling methods and logic to be used as analytical instruments and techniques to organize the process in cognition in order to define the path of thought as an *organic enveloping* of decision-choice actions. The objective is to simplify the complexity of knowing through some logical filling under some rationality. The analytical position in this theory of the knowledge square is that the whole theory of knowing is decision-choice determined from defective information structure. It is under the conditions of decision-choice determination that cognition traverses over the epistemological space to transform information into claimed knowledge which is always conditional but not absolute. The conditionality specifies the applicable areas of claims and the degrees of risk associated with the epistemic claims. The rise of this conditionality and the factors that motivate it are made explicit in the theory of the knowledge square and shown to relate to quality and quantity of information.

K.K. Dompere: The Theory of the Knowledge Square, STUDFUZZ 289, pp. 1–12.
springerlink.com                    © Springer-Verlag Berlin Heidelberg 2013

The theory also offers us an approach to conceptualize the importance of and the relative positions between the *classical paradigm* and the *fuzzy paradigm* as we view their logics and their corresponding mathematics that are used to guide reasoning in complex systems of universality and particularities in all levels of philosophy, physical, chemical, biological and social sciences and their mutual interactions under conditions of *time, quality and quantity*. In the whole process of knowledge construction and reduction, one encounters different problems of clear understanding of linguistic representations as well as interpretations of their meanings irrespective as to whether one is working in humanistic and non-humanistic systems. The absence of clear understanding and unquestionable interpretation sometimes lead to claims of contradictions and paradoxes in knowledge acceptances as they relate to human experience and its management. The discussions and the geometry of Aristotelian logical opposites in the knowledge production process will be related to geometrically pyramidal structures in fuzzy logic and its opposites. In this respect, the meta-theory will bring into focus the sources of uncertainty and risk over the epistemological path. We shall specify the sets of conditions that give rise to the possible, the probable, the possibly probable, and the probably possible that lead to the theories of uncertainty, possibility, probability and inexact probability.

## 1.1 Ponts of Entry and Departure of the Theory of the Knowledge Square

The general foundation of our knowledge construction is shaky with some serious weak points such that the superstructure of our knowledge is being constructed without clear understanding of the pillars that support the house of knowledge that is being erected, and under continuously perpetual improvements and beautification. For example, there are some intellectual claims of existence of pure mathematics and exact science with some disrespect to distinguishing themselves from what they refer to as applied mathematics and inexact sciences in the domain of production of human knowledge. Interestingly, new knowledge seekers enter into the space of the knowledge search and build on the top of the knowledge house without examining the strength of its foundations and the pillars that support it. The contemporary information science, environmental science, clamatological science, medical and biological sciences and economics seem to suggest that this position of distinction between exact and inexact sciences cannot be held without question. In the process of our continual building and improving the house of knowledge, some initial foundational assumptions have come to be taken as natural and self-evident truths rather than as analytical conveniences in simplification and initialization of the building process.

As such, the role of decision-choice process and its rationality in the knowledge-construction enterprise have been lost at higher levels of abstraction. These higher levels of abstraction are sometimes taken as justified by the assumed self-evident truths from the initial conditions. Our global knowledge structure has come to be accepted as *real* and completely believable to the neglect of the recognition that it is the building result of human cognitive construct with all its difficulties and flaws that need continual reexamination for cognitive tuning. Similarly, the logic of its construct has come to be accepted as the ultimate path to the knowledge house by the knowledge seekers including both the old and the new comers. The uncompromising belief in such an ultimate logical path becomes ideological which constrains the development of other possible logical categories in the possibility space. In this way, the belief imprisons imaginations and creative intuition within the wall of its ideology and refuses to allow cognitive freedom outside it. This seems to apply to all sectors of knowledge production, the process of teaching and learning at higher levels of education, research and knowledge search where one path of methodological reasoning is claimed to be the ultimate to a knowledge discovery. The belief-system is protected by those who have something to loose. A need arises to relax some of the initial assumptions that define the classical framework as the domain of rational knowledge development expands to meet new challenges of human decision-choice actions.

The human knowledge construction operates through the dynamics of creative destruction where some old knowledge items, one time accepted as credible, become discredited and new ones are created in their places. This is the constructive-destructive duality with a continuum in the knowledge production process without end. Thus the development of human cognition runs into limits and the construction of the global knowledge house runs into material constraints. New experiences allow us to relax the material constraints and sometimes throw the existing logic into a flux. The foundation of the house of our knowledge construction is under continual conflicts in a *true-false duality* and *quality-quantity* changes. These conflicts are useful in revealing points of critical stresses and the understanding of the conflicts at higher levels of reasoning among different knowledge sectors. The continual creation of these conflicts has two aggregate sources. One source is from the logic (methodology) of its construct. The other is from new experiences (changing information structure) that challenge our justified beliefs and the accepted knowledge elements. The aggregate epistemic effect is to continually reexamine the foundational path of our knowledge production itself and the strength of its analytical pillars that must be related to its very foundation; the foundation of thinking and knowing where the knowledge production is traversing on the enveloping path of the resultant forces of construction-destruction duality in a continuum.

It is here that *the theory of the knowledge square* takes as its *point of entry*. It is meant to present a framework to examine the integrated key concepts and ideas of the building foundation of the house of the global knowledge. The concepts and ideas include measurements, certainty, vagueness, exactness, inexactness, particular, universal, symbolic representations, interpretations, laws of reason and many others. The essential utility here is the examination of human cognition into the forces at work in understanding nature and society, and how such understanding will provide a source of continual improvement in our justified belief system in using this understanding to examine claims of knowledge that offer us a useful toolbox to further understand the interactive forces of nature and society at work, and hence to manage society for the advantage of humans. The continual reexamination is demanded by the very nature of our knowledge production as self-correction and self-improvement system.

Traditionally, the forces of nature and society have been separated in our quest for understanding and knowledge without the understanding that nature imposes its will on the society while society imposes transformability on nature within defined boundaries of human capacity which expands with increasing knowledge. The forces of nature are considered beyond human intervention while the forces of society are subject to human creation and intervention. The methods of inquiry into these forces, as well their understanding, are also separated with less and lesser emphasis on the role that decision-choice processes play, particularly, in the knowledge sectors involving natural processes. It is only in some knowledge areas of social processes that human decision-choice activities in knowing are studied as part of the social process itself. It is not surprising that some subject areas of the knowledge-production process, such as mathematics, statistics and symbolic logic are studied, and presented as abstract systems devoid of human decision-choice actions and hence completely separated from the linguistic structure of human interactions, information and existence.

The activities in these abstract systems are presented as games of thought within each system that is guided by strictly strategic rules of reasoning which then provide us with true-false configurations. Furthermore, the roles of institutions and those of social belief systems that either facilitate or restrain the domain of knowledge search are neglected. The end result has been a situation where some knowledge sub-structures have come to loose the basic notion that they are human constructs whose legitimacies are derived from the social system itself through collective decision-choice actions with vagueness, imperfections and inexactness. These qualitative elements of vagueness, logical imperfections and linguistic inexactness result in a situation where any claim of knowledge purity must be in degrees of perfection which are decision-choice determined.

Our global knowledge system is a collection of cognitive models of interactions between ontology and epistemology with phrases of "what there is", "what there is not", "what ought to be", "what could not be" and many similar linguistic characterizations of the knowable and unknowable phenomena through epistemic actions. The process of knowable and unknowable, with differential belief systems for their justifications, challenges human understanding of forces that are operating in the nature and society to shape their qualitative and quantitative structures. Our accepted knowledge is not the reality of these forces at work. It is simply an epistemic model of what our cognition and thought project. It is the recognition of, and the emphasis on the role of human decision-choice action in cognition, and the internalization of the role of the decision-choice processes in cognition, that the theory of the knowledge square presents a *point of departure* from the accepted tradition of knowledge production.

It is in the search for a path to internalize the human decision-choice process, with its rationality, ambiguities in its language and faults of reasoning into the knowledge production process, that establishes an *entry point* for the theory of the knowledge square as an approach to understand claims and counter claims in science, non-science, exact science and inexact science. These points of entry and departure have been made possible from the developments of fuzzy paradigm whose laws of thought allow us to replace the classical law of excluded middle with the fuzzy law of continuum in order to internalize the decision-choice actions and deal with the complexity and synergetic structures. The approach taken here is not only that ontology and epistemology are seen in a unitary form, but that the process of knowing is separated from the methods of knowing and reconnected under the general unity of epistemics.

The theory of the knowledge square will allow us to place in proper interrelated epistemic modes, the dynamics of the essential components of methods and tools of knowledge production and the evolving path of the knowledge system. It will, further, allow us to demonstrate the unity of knowledge sectors where the study of natural sciences is embedded in the study of social sciences and the gains in the natural sciences reinforce the integrity of the study of the social sciences. The study of the social sciences is not embedded in the study of the natural science. From the viewpoint of social formation and in relation to cognition, there is no natural science without social science. In fact, a claim can be made that natural science is embedded in social science and that a progress in the study of categories of the natural sciences is a reflection in the progress in the study of categories of the social sciences. Areas of natural science and their studies are just as exact and certain to the extent to which social sciences and their studies are exact and certain through improved efficiencies of social management.

The interesting thing is that humans are part of nature whose forces are under human cognition and whose process can not be separated from the society. The degrees of success of this cognition depend on the established social institutions that contain the social forces which are in turn under cognition and further institutional formation. In both cases of the study of forces of nature and society, cognitive agents interact and influence the subject of study, the problem selection and the outcomes of cognition, and all these take place always under a given social institutional configuration. The study of natural phenomena, just as the study of social phenomena, is a social activity governed by the dynamics of society as a self-regulating, self-correcting and self-exited system under continual quality-quantity transformations.

The nature of categories of the natural sciences in terms of exactness and the efficiency of the study of the forces of nature, and the conclusions derived from the studies are reflections of the developmental complexities of the society, its institutions and methods of knowing. The developmental complexities of society are further complicated by the need to understand the equations of motion that govern their qualitative and quantitative dynamics and their interactions with non-social processes. The knowledge about natural forces is also the knowledge about social forces and vice versa. In an essence, the knowledge about society and the knowledge about nature form an inseparable unit that reflects the existence of polarity, where each pole presents a duality [R2.9] [R11.20], [R11.21] [R14.29] [R14.30] that allows the understanding of the dynamics of the knowledge production.

It is also from this angle of the concepts of polarity and duality, as reasoning instruments in the knowledge production process, that one may see the *unity of science* and a search for global systemicity and synergetics towards the knowledge production, where the geometry of thinking allows us to integrate the science of synergetics with complexity theory towards a universal principle of the knowledge production process. Unfortunately, our knowledge production system cannot take claim to any principle of universal validity. This seems to characterize both the natural and social sciences. The division of knowledge areas into natural, behavioral and social sciences with further sub-fields must be viewed as merely an analytical convenience of the enterprise of the knowledge production on the basis of decision-choice process with specified rationality and divisions of labor. Such divisions may be viewed in terms of sequential problem solving that allows social division of labor and specialization. From the point of view of societies, there is one knowledge house with many interconnected rooms and more rooms to be added for continual expansion and beautification. The question is: how is this house being constructed and how strong are its supporting pillars?

It is the same decision-choice process that has artificially created exact and inexact sciences and maintains them with ideological vengeance and rigidity. It is also the same decision-choice process that will bring them into unity by dissolving the ideological walls that separate them. Our cognitive reflection is that exactness and inexactness exist in duality in the same way as, vagueness and clarity exist in duality where the understanding of the internal conflicts in their dualities provides us with conditions of categorial conversion from inexact to exact through a substitution-transformation process of logical dynamics in continuum. A question arises as to whether we consider exactness as *primary logical category* from which inexactness emerges as *derived logical category* or the other way around where inexactness constitutes the primary logical category from which exactness emerges as a derived logical category. By the use of instruments of the development in the theory of the knowledge square, we shall present a position that there are no differences between natural and social sciences and between exact and inexact sciences at the level of the path of knowing.

The generally accepted social position and in the academy is that social science medical sciences, environmental sciences and behavioral sciences are inexact [R8.53][R8.54] [R14.88] [R14.97] and that natural science is exact. This is an ideology and we shall demonstrate that this ideology is created on a weak foundation and unsustainable assumption. The same characteristics that are used to separate science into exact and inexact sciences will be shown to unite them as a combinatorial knowledge-production unity. The same elements used to create accusations of inexactness against social and behavioral sciences will be used to strengthen them and then used as accusations to strip off conditions that define exactness of natural sciences. To do this, we must establish a *universal principle* of knowledge search and knowledge production that holds for all areas of cognition. Let us turn our attention to the structure and form of the epistemic process that will establish the universal principle and its conditions for knowledge production. The knowledge square provides us with a cognitive geometry of the universal sequence for the knowledge search, and this universal sequence will constitute the universal principle for knowledge development in the epistemological space. There is another thing in that the proponents of exactness in the knowledge production fail to account for the simultaneous existence of qualitative and quantitative characteristics of information characteristics of epistemic objects.

## 1.2 The Knowledge Square

What is the knowledge square and how is it related to the understanding of cognition and the establishment of the universal principle for knowledge search in the enterprise of knowledge production? Alternatively stated: what is the relationship between the knowledge square and the theory of knowledge? Our task which is a simple but difficult one is to define and specify the epistemic foundation of knowledge and how the evaluation of such a foundation is established by decision-choice processes of cognitive agents as part of the nature and society under epistemic processes. We shall then relate the theory of the knowledge square to the traditional view on the theory of knowledge that is decision-choice based. Within this decision-choice process, the knowledge square is established and composed of four epistemic building blocks with cognitive connecting cords that strengthen the foundation of the knowledge enterprise. These blocks are a) the potential space, b) the possibility space, c) the probability space, and d) the space of the actual. These spaces are knitted together by decision-choice actions of cognitive agents in the search for "what there is", its knowability, explainability of its behavior, and the understanding of its substitution-transformation process. Together these blocks constitute the pillars of epistemology in a form that begins with the general and works its way to the specifics from the ontological space. In an essence, the development of the theory of the knowledge square is embedded in the general synergetic structures as geometry of cognition. The building blocks and their cognitive connecting cords as we will discuss are geometrically presented in Figure 1.2.1.

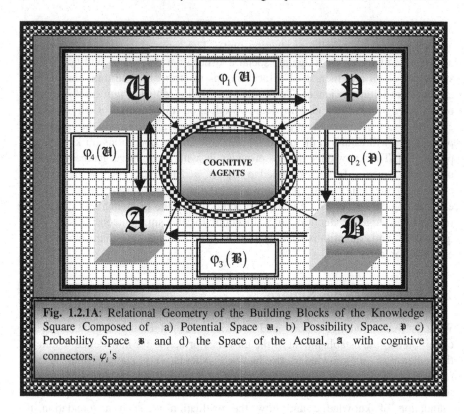

**Fig. 1.2.1A**: Relational Geometry of the Building Blocks of the Knowledge Square Composed of   a) Potential Space $\mathfrak{U}$, b) Possibility Space, $\mathfrak{P}$ c) Probability Space $\mathfrak{B}$ and d) the Space of the Actual, $\mathfrak{A}$ with cognitive connectors, $\varphi_i$'s

Each of the building blocks is a family of categories. The cognitive connectors are logical processes that allow substitution-transformation processes to occur between blocks and within blocks. These connectors are the works of information and logics with assumptions that define the environments of categorial conversions whose dynamics are governed by conflicting forces in dualities that are embedded in languages of thought. At the center of the knowledge square are cognitive agents whose decision-choice actions determine what items are knowledge ones and what items are not, as we journey from the *knowledge potential space* to the space of cognitive reality, and as the contents of the knowledge house are being determined for epistemic filling. We shall deal in conceptual details and separate the essence of these blocks and establish their relational structures to the foundation of the knowledge enterprise. With the instruments that will emerge out of the theoretical framework of the knowledge square, we shall examine science, exact science and inexact science in order to establish their mutual similarities and differences. It will become clear that the path of the knowledge square is invariant to areas of knowledge search no matter how much segmentation is created and how many segments are imposed on the global area of knowing. We shall then examine intra-block and inter-block logical transformations that define the path of cognition as the universal principle of knowledge search and development in the epistemological space that is linked to the ontological space.

## 1.3   The Knowledge Square and the Concept of the Potential Space 𝔘

In the theory of the knowledge square the *potential space* 𝔲 is used to initialize the knowledge production process. It is useful to think of it as a cognitive potential space that contains all the elements of the universe called the *universal object set*. The universal object set is made up of elements, states and processes including all cognitive agents such as humans. It is potential in the sense that the elements may be exactly or inexactly knowable. These elements have essence and identity that are defined by *characteristic sets*. The characteristic sets belong to a family of characteristic sets where each set may be used to distinguish, classify and name the objects into categories by cognitive agents in a given language. The universal object set may be represented as, $\Omega$ with generic element $\omega \in \Omega$. The potential space for knowing may also be viewed as the ontological space. In other words, the development of the knowledge square is to search for conditions that distinguish the ontological space from the epistemological space with a further search for conditions that unite them. In this process, a number of definitions are required.

### Definition 1.3.1: The Cognitive Potential

Cognitive potentiality is a schematic foundation of the knowledge square that initializes the conceptual system for the epistemic process in order to connect the epistemic space to the ontological space.

### Definition 1.3.2: The Universal Object Set, $\Omega$

The universal object set is the collection of all objects, states and processes that present potentials of being known by all elements in the universal object set. That is

$$\Omega = \left\{ x \mid x \in 𝔲 \text{ and } x \text{ is potentially knowable} \right\}$$

### Definition 1.3.3: The Potential Space, 𝔲

The potential space is an infinite collection of families of categories in the universal object set and is infinitely closed under the substitution-transformation process. It is also the ontological space.

The set $\Omega$ is defined in the potential space, 𝔲 and contains infinite elements that are potentially identifiable and knowable through cognition by cognitive agents who are also part of the universal object set. Nothing in the universe is outside the universal object set or the ontological space. The processes of identification and knowing begin with awareness and visionary activities that help to establish the perceived characteristics and their essences. We may note that 𝔲 is the potential space while $\Omega$ is a family of essential categories, each of which contains common elements and common defining characteristics as part of characteristic set $\mathbb{X}$ with generic element $x \in \mathbb{X}$. For example, we may mention the category of living things with further categorization into cognitive and non-cognitive living things. The potential space $\mathfrak{U}$ is also the ontological space and the universal object set is a collection of *ontological elements* such that $\Omega \subseteq \mathfrak{U}$ .

### Definition 1.3.4: Epistemological Space, $\mathfrak{E}$

The epistemological space is the union of the possibility space, the probability space and the space of cognitive actual such that $\mathfrak{E} = \{ \epsilon \mid \epsilon \in (\mathfrak{P} \cup \mathfrak{B} \cup \mathfrak{A} )\}$. The symbol $\epsilon$ is an epistemic element where $\mathfrak{E}$ is a collection of all epistemic elements

The potential space, $\mathfrak{U}$ composed of the elements in the universal object set, $\Omega$ is taken to be that which actually exists in our universal system. Its existence, and the elements contained in it are independent of the awareness of its elements whether they are cognitive or non-cognitive agents. In other words, they will exist whether one is aware of them or not. It is the potential in the sense of knowability through cognition. The potential space, as an ontological space, is linked to the epistemological space through the processes of knowing where the epistemological space is the collection of all epistemic items and categories. The process of knowing, where the elements of the ontological space are to be discovered as knowledge items for storage by cognitive agents, begins with the awareness of the defining characteristics of an element in the potential space through identification, naming and category formation in the form $\mathbb{C}$ with a generic element $(x, \omega) \in \mathbb{C}$ such that $x \in \mathbb{X}$ and $\omega \in \Omega$. The characteristic identification, naming, and category formation are the works of information transmission and interpretations as sense data that operate through cognitive agents, whose very existence is also part of the universal object set. It is the development of the *universal principle of knowing* that is the concern and the subject matter of the theory of the knowledge square as a geometry of thinking.

To clarify a tendency that may bring some confusion, the theory of the knowledge square postulates that the objects in the potential space are *natural actual* and the characteristics are their naturally defining attributes that exist independently of the awareness of the objects in the potential space. The set of characteristics that establishes identity, difference and similarity of objects in the universal object set may be referred to as *natural characteristics*. The category to which each object belongs in the potential space may also be referred to as *natural category*. There is, however, a cognitive process to establish parallel elements that can be compared to the natural elements for knowledge acceptance in the process of knowing and knowledge production. These objects that come to be known through cognition are classified as *epistemic actual* when the elements are claimed to be known. Thus corresponding to natural categories we have *epistemic categories*, corresponding to natural characteristic set we have *epistemic characteristic set* and corresponding to natural actual we have *epistemic actual,* all of which are under the process of cognition.

The task of the knowledge construction is to show that the epistemic characteristic set is reasonably identical or have sufficiently close resemblance to the natural characteristic set for each element, the epistemic category formation is identical to the natural category formation and that the epistemic actual corresponds to an aspect of the natural actual. In other words, we must show the isomorphism between the epistemic actual and the natural actual through cognition. Alternatively, we must show that the space of epistemic actual is contained in the ontological space in that $(\mathfrak{a} \subset \mathcal{U})$ but $(\mathcal{U} \not\subset \mathfrak{a})$ at all times. The process of showing this isomorphic relationship between epistemic elements and natural elements is decision-choice driven where inexactness, vagueness and approximations are attributes of its foundation in any language and reasoning for searches in both natural and non-natural spaces.

The natural condition is ontological and the process of knowing is epistemological. Thus, there is objective information that defines the objective characteristic set, the family of natural categories, and potential space and its elements and categories. The question that the theory of the knowledge square seeks to answer is what is the process and universal principle through which cognitive agents acquire knowledge about aspects of the potential and the universal object set to which they belong? Another way to see the problem is through the question: what is the path or the geometry of thinking for establishing the isomorphism between cognitive elements and natural elements or between epistemic elements and ontological elements and what is the universal principle for establishing the path of cognition?

There are a number of key concepts that have been introduced in this analytic framework. They need clarification in terms of their meaning and content. Definitions and explications are required for a) the potential space, b) the universal object set, c) the natural characteristic set, d) the universal characteristic set and e) the natural category. These are the initial and basic building blocks of the theory of the knowledge square as it relates to the understanding of the knowledge production and enterprise of knowledge. They are taken as axioms that initialize the knowledge construction process regarding knowability and explanability. They allow us to assert the ontological conditions of the universe and its elements [R2.9, pp.54-71] [R2.2.11]. They further allow us to derive the epistemological conditions of knowing. The conceptual system of the potential space is composed of postulates and axioms of existence of the universal system whose elements are to be known, through experience, and whose behaviors are to be explained through reasoning. The existence of the universal object set is taken to be independent of the existence of the awareness of its constituent elements whether these elements are cognitive agents or not. Alternatively, the ontological space is conceptually separated from the epistemological space and then reunited by thought and reasoning for the knowledge production.

The question facing cognitive actions is simply how the elements in the universal object set come to know their own existence and the existence of other elements. It may be pointed out, at this juncture, that care must be exercised in conceptualizing the universal object set as defining the nature of the universe without confusing it with the classic debate on the relationship between the universal and the particular [R8.62][R14.96] [R14.97]. In this monograph, the universal is the set of particular objects where the collection of all particular objects with similarly defined characteristics constitutes a category. The universal object set is a mega-category in the sense that it is the family of families of particular categories. The concepts have nothing to do with universality and particularity of propositions, statement and hypothesis.

The objects may be elements of states or processes or both. The conceptual system that we are advancing is on the *principle of duality* where there is no universal without the particular and there is no particular without the universal. The principle of duality is supported by the principle of *continuum* in the geometry of thinking where the particular is linked to the universal and vice versa. The universality and particularity mutually define themselves in identity and their existence in unity with continuum such that the process of reductionism allows the creation of quantum units. The concepts of the universal and the particular are in relation to objects of states and processes that establish the potential space; they are not in relation to propositions, truth and knowledge acceptance. The potential space is defined by *what there is* not a cognitive construct. It may be viewed as the ontological space that is the collection of *all what there is*. Let us connect the ontological space to the process of knowing through the geometry of thinking from the possibility space.

# Chapter 2
# The Theory of the Knowledge Square and the Concept of the Possibility Space, $\mathfrak{P}$

We now turn our attention to the *possibility space* $\mathfrak{P}$ as the second building block of the knowledge square. The possibility space is a cognitive connector between the potential space, $\mathfrak{U}$ and the space of the *natural actual*. It is an element of continuity of the epistemic chain of knowing and explaining where the potential space and its contents, relative to cognition, are taken to exist but unknown to cognitive agents as true ontological elements. The potential space is the initialization of activities of cognitive agents and their knowledge production process where knowledge and awareness of the elements must be abstracted by a process. The possibility space is made up of elements which are possible candidates for knowing from the potential space. It is cognitive construct in terms of ontology and epistemology. It constitutes an initial bridge between ontology and epistemology (the words epistemic and cognitive are interchangeably used in a number of occasions). The question, therefore, is: how are the concepts of possibility and possibility space defined, how are they connected and how are they related to the uncertainties and the process of knowing?

## 2.1 The Concepts of Possibility and Possibility Space

Let us turn our attention to the concept of the possibility space. This concept is linked to the concept of the possibility and what constitutes possible in the language structure and how the possible is related to uncertainties in cognition. The form that we want to construct requires definitions and explications of these concepts to which we turn our attention.

### Definition 2.1.1: Possibility

Possibility in the conceptual scheme of the knowledge square is that which is cognitively abstracted from the universal space or the universal object set through acquaintances, senses and vision with justified degree of belief that the element may be a knowledge item. The impossibility is that which has zero justified degree of belief.

K.K. Dompere: The Theory of the Knowledge Square, STUDFUZZ 289, pp. 13–41.
springerlink.com

## Definition: 2.1.2:` Possibility Space, $\mathfrak{P}$

The possibility space defines a conceptual category that provides a linkage between the natural elements of the universal object set and human imagination, intuition and conscious existence which allow a construct of that which is epistemologically believed to be informationally possible as knowledge within the universal object set (the potential space) and with differential degrees of *possibilistic uncertainty* associated with the elements. The collection of the conceived elements in the universal object set that are moved into a space whose elements are believed to be possibly knowable constitutes the possibility space $\mathfrak{P} = \left\{ \left(x, \mu_{\mathfrak{P}}(x)\right) \mid x \in \mathfrak{U} \text{ with } \mu_{\mathfrak{P}}(x) \in (0.1] \right\}$ where $\mu_{\mathfrak{P}}(x)$ is an index of possibilistic belief in the knowledge existence.

The imagination, intuition and conscious reflections generate vagueness in ideas and ambiguities in thought and communication called *fuzzy uncertainty* which produces *possibilistic belief system* and *possibilistic risk system* in the knowledge production and decision-choice processes for assessing the knowledge content at all stages of knowing. The possibility space comes to us as a schedule with elements and corresponding distributions of degrees of possibilistic belief in the form $\left(x, \mu_{\mathfrak{P}}(x)\right) \in \mathfrak{P}$. The cognitive instruments for the construct of the possibility space are obtained by imagination, intuition, information signals, principle of conscious existence and a system of beliefs. The outcomes in the possibility space, while useful, do not constitute knowledge. They merely point to the possibilities of knowledge. They help to identify the possible elements that require expenditure of cognitive labor and hence provide us with the instruments that allow us to construct the possibility space from the potential space. That which presents itself to the consciousness or reflected in the intuition is merely a possibility to an epistemic reality. It may also be an epistemic mirage. The elements in the possibility space may be classified as a sub-level of observation of pseudo-empirical data that are individually specific which has no truth-value except degrees of possibilistic beliefs.

Intuition, imagination and conscious activities provide us with a path toward the construction and transformation of the universal object set, $\Omega$ into the possibility set, $\mathfrak{P}$ , where $\mathfrak{P} \subset \Omega \subseteq \mathfrak{U} \supseteq \Omega$. This is on the epistemic path that allows us to construct a set of possible epistemic elements that may be considered as knowable in the sense of *what there is*. The general knowledge acquisition by cognitive agents is a never-ending process in a self-correcting system where each knowledge item, after acquisition, reveals a new item of unknown character. It is socially, individually and collectively cognitive activity. The institutional provision of its framework provides cognitive agents with social interaction toward a common goal, the goal of knowing through epistemic reflections of multitudes of phenomena for varying individual and collective needs. In this process, the concept of the possibility space is one of the organizing tools towards the understanding of human cognition and knowledge acceptance as deliberate intellectual activities. The philosophical debate of actuality and possibility on one hand and intuitionism and formalism on the other does not arise here. It is considered as irrelevant when one accepts that the elements of the universal object set, composed of objects, processes and states, are relational in existence, and so also the knowledge construction towards knowing ([R8.59] [R8.56] [R14.9] [R14.14], [R14.23], [R14.52] [R14.97]).

It is at this point of representation of information in the path of knowing that analog and digital processes enter into the knowledge production process. The question is simply whether reality is analog or digital; continuum or discrete; or inexact or exact? The answer provided in this discussion reflects a position that the concept of reality must be dichotomized into ontological and epistemological where *ontological reality* may be different from *epistemological reality*. The epistemological reality is a model representation of ontological reality as such they may stand to differ. It is a position in this discussion that ontological reality is analog and exact while the epistemological reality may be digital or analog or both by epistemic construct. It will be argued later in the monograph that whether one subscribes to epistemological reality as analog or digital will reflect on the methodological approach of information representation, methods of analysis and techniques of knowing. The nature of epistemic reality that is held will reflect the initial assumptions about the structure and the properties of *what there is* in the ontological space. There is a relationship between the ontological and epistemic realities where one is prior and the other is posterior. When one takes the ontological reality as a primary category and that this primary category is analog and exact, then the problem is to abstract the conditions of their relationship and how the epistemological reality acquires digital representation. The relational structure is shown in Figure 2.1.1.

A possibility reflects itself as intuitive and imaginative actual with subjective belief justification but without a demonstration of its validity by cognitive agents. It is a step in verifying the properties of the epistemic actual against that of the natural actual for their acceptance into the general knowledge bag of humanity. A possibility space, therefore, is a *possibility complex* composing of a family of sets of possible elements even given the same phenomenon. It is a family of families of elements of intuition, imagination and contemplation. It is here that the concept of *possible worlds* may enter into our knowledge production system. It is a stage in the knowledge-production process that is subjectively reflected in intuitive elements in the universal object set. It is a part of the connecting path of the geometry of thinking to establish an isomorphism between some ontological elements and epistemological elements [R12]. It is also here that the processes of sequential and digital methods of representation, analysis and problem solving become tools for knowing. The human cognition works within a constraint set such that cognitive agents do not generally work in the space of analog but rather in the space of digital for simplicity toward knowing. Analog is complex and digital is simpler. The mind likes to work with simplicity and avoid complexities whenever possible.

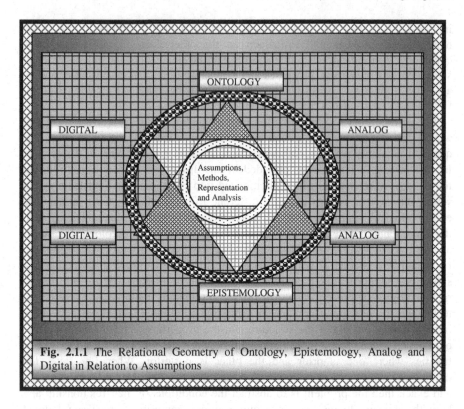

**Fig. 2.1.1** The Relational Geometry of Ontology, Epistemology, Analog and Digital in Relation to Assumptions

## 2.2   The Possibility Space, the Spaces of Acquantance and Language

The connection between ontology and epistemology begins from experiential acquaintance and proceeds with naming, definition, meaning and communications of the acquainted items. We shall refer to the process of naming as *nominalism* that requires representation by either symbols or words in a given language. We shall refer to the process of conceptual contenting as *definition* while the process of communication, we shall refer to as a *language* that operates through oral or written symbols in order to describe, distinguish, classify and catalogue objects, concepts and experiences into categories for an individual and group distinctions. The space of nominalism is digital while the space of language has an analog beginning. The symbols appear to us as words which come to represent those objects, concepts, processes and notions in the universal object set to which intuitive and imaginative experiences reveal themselves to initial cognition. Concepts are formed around the symbols of objects where the symbols are defined for cognitive processing. Not all objects in the universal object set have names. It is only the objects in the universal object set that have acquaintances that are named. The implication is that all elements in the possibility space have or will have names and linguistic meaning. Names are digital while meanings are analog.

Any element in the universal object set that has human acquaintance is assigned a word or a symbol and since these symbols are not unique from the space of acquaintance, it becomes necessary to fix their meanings and contents by definition and explication in a given language depending on the uses to which they are to be put and the role that they will play in human understanding in the universe of knowing and communication of understanding among cognitive agents. There are two initial problems associated with any word in the space of acquaintance given the object of acquaintance. They are the object's name and the language of its representation and residence. The definition of a word or a symbol must correspond to the object of acquaintance or its derivative thereof, where the object may be an element, a process or a state. There is no law of nature that an exact correspondence can be created between an ontological element and an epistemic element in the process of cognition. At the very best, we may have cognitive approximations in experiential data, linguistic representation and communications.

The processes of naming, defining by symbolic representation and the development of languages were not initially conscious efforts. They came to us through epistemic evolutions with continual cognitive refinements. Language and knowledge production are inseparable. They are also self-refining and self-improving systems as unconsciously social constructs. They became conscious activities at a later period of human development and at the time when further development demanded clearer understanding of the elements in the space of acquaintances in order to communicate the objects of acquaintance as something dependable to either use or fear or share. Language emerges as representation and communication of information. Communication and understanding through the linguistic medium demand clarity of words, symbols and their meanings in the language of cognition as a carrier of that which is acquainted. Similarly, the thought that is being formed around the acquainted became necessary. Another way of looking at the process is that cognition begins as analog whose development fades into discreteness or digital, where the fundamental analog and discreteness may be related to the fundamental notions of exactness and inexactness.

The need for communication among cognitive agents from human experiences in the space of acquaintance (defined in terms of intuition, contemplation and sense data) in the universal space, not only gave rise to language, but the language gave rise to the capacity to form ideas and construct thoughts on the basis of objects of acquaintance, their description, behavior and explanation. All these cannot be separated from conditions of economic production and human survival. The space of acquaintance is, thus, a subspace of the potential space for knowability. It projects nothing but speculations couched in thoughts of opinions and subjective reflections about some elements in the universal objet set that contains the elements of the space of acquaintance. The space of acquaintance is not the knowledge space; and the opinions and subjective reflections about the intuitive elements are not knowledge elements but elements for knowing called *epistemic elements* that may be made possible by operations on words in a language and formation of thought for knowledge verification.

There is nothing unique about language and thought that can take claim to exactness relative to natural elements in the universal object set except by cognitive imposition and acceptance through decision-choice actions. In fact, linguistic symbols (words) on the basis of which ideas are formed are ambiguous in representation, vague in meaning and approximation in interpretation whose crisp understanding and exactness of meaning are the results of human decision-choice actions. The spoken language represents a combination of sound symbols of words that are coded with information for interpretation, processing and understanding by members of the linguistic group in accord with the rules of combination for knowledge development through judgment and decision making. The same sound symbols may be coded differently with different information contents and with different interpretation by members in another linguistic group also in accord with rules and decision making. The written language represents inscriptions of written symbols of sound symbols of words coded with information content for interpretation, processing and understanding by the linguistic group in accord with the implied grammar for knowledge development through decision making. The symbolic nature of the written language varies over linguistic groups even though they may carry the same information. In this way, information, knowledge and decision-choice processes are inseparable and reside in a unity where the transformation of information to justified knowledge is decision-choice determined through the methods of analog or digital as presented in Figure 2.2.1.

In general, language is the foundation of communicating acquaintances, development and exchange of ideas, and propagation of thought which then motivates thinking and further refinement in a never-ending process of the knowledge search. The thinking relates things in the space of acquaintance through elaboration and development of multiplicity of concrete and abstract connections. In this respect, it is obvious that any language is arbitrary with its own character of defined conventions in symbolism of representation, and that all languages carry with them the qualitative properties of vagueness, inexactness, ambiguities and approximations in meaning and thought. Every language has a character and form that carry the culture and forms of thought. It is this qualitative arbitrariness that allows different language formations to be established. It is also the formation of categories and rules of associations that provide channels of translation from one language to another and where the translations are sensitive to the cultural confines of languages.

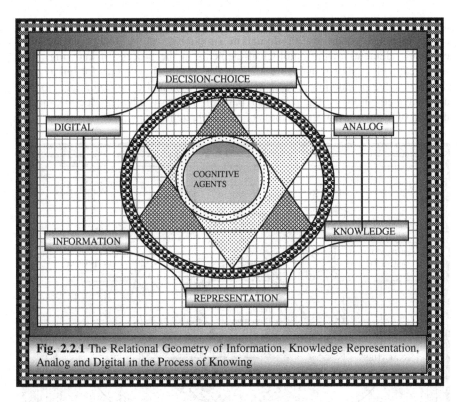

**Fig. 2.2.1** The Relational Geometry of Information, Knowledge Representation, Analog and Digital in the Process of Knowing

The characteristics and categories in the universal object set are the same for all language formations but differently named as well as differently related to create concrete and abstract thoughts. Furthermore, acquaintances may be different for different human species in different language representations. The ideas and thoughts are carried by languages and are exact to the extent to which the languages can be made exact for any given culture. Ideas are formed with words and linguistic connectors in a complex system of organization that allows a linkage to be made of some elements in the universal object set to the space of acquaintance in either conscious or unconscious state. The acceptance of exactness of a language, in addition to the ideas and thoughts it carries, is dependent on the conventions which come to define the grammar and the laws of thought of that language. In this respect, vocabulary of any language, rules of its grammar and syntax are established by conventions that are fixed by collective needs of the social formation including the evolving culture.

The vagueness, inexactness and approximations in languages are to all linguistic forms general but not specific to subspaces of acquaintances. They contain qualitative and quantitative dispositions of elements that define their identities. The acquaintances are related to cognitive agents though a complex system of information signals and communication. It is this vagueness character of languages that generates increasing complexities in language and for language translations. The vagueness is generated by the conditions of the fundamental analog or continuum that imposes requirements of decision-choice actions to create assumed conditions of digital in simplicity, as we seek to relate the epistemic elements to the ontological elements in the process of knowing. The analytical path is shown in Figure 2.2.2 as the relational geometry for initializing the knowledge production process and how any assumed condition relates to the adopted methodology of a particular path of the knowledge production in the domain of exactness and inexactness of epistemic items.

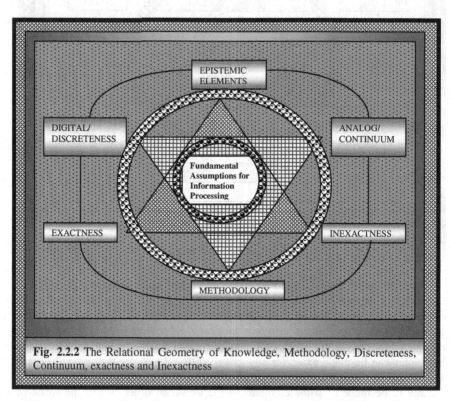

**Fig. 2.2.2** The Relational Geometry of Knowledge, Methodology, Discreteness, Continuum, exactness and Inexactness

## 2.3   From the Space of Acquaintance to the Possibility Space

The space of acquaintance is formed between complex interactions of cognitive agents and other elements in the universal object set through information transmissions and interpretations by the sending-receiving processes [R8.18] [R8.50] [R11.19]. Each element in the universal object set, $\Omega$, is identified with a set of relational characteristics, $\mathbb{X}$ that defines its identity and allows a classificatory process to be undertaken for the expansion of linguistic vocabulary, the formation of ideas and thought developments from the elements of the space of acquaintance. The collection of the elements with the same relational characteristics from the space of acquaintance is called a *category* (for example, people, rivers, animals, ideas, propositions, atoms and other.). The collection of the categories of the epistemic elements together with anticipations of knowledge constitutes a family of classes of categories called the *possibility space* under cognitive action.

It is at this juncture that the laws of thought become necessary and essential in forming the categories. The formation of categories is not different from the formation of class, sets and groups whose members have specified characteristics that are decision-choice defined in the possibility space. The formation of these categories cannot be done arbitrarily. It must follow some socially defined convention of the language of the process. It is here that the classical Aristotelian laws of thought enter [R14.58], [R14.66] [R14.97] [R14.100] into the formation of sets, categories, groups and the families as well as information processing and knowledge construction. It is also here that a distinction is made between the *fundamental discreteness* and the *fundamental continuum* in the initialization of the knowledge search process. The Aristotelian laws of thought initialize the knowledge search with the fundamental discreteness and present a rigid thinking in *dualism* where the elements in the space of acquaintance are viewed in {0, 1} sequence.

These Aristotelian laws of thought in their essential form may be stated as *all propositions are either true or false but not both* which is carried on to mathematico-scientific reasoning where the claim of exactness and crisp categories imposes a rigidity of thinking, such that the elements in the space of acquaintance belong to a category or not and their symbolic representations are exact in meaning. It is here that mathematical language and thought are claimed to be exact and hence removed from the subjectivity of judgment as human action. The Aristotelian laws of thought, as applied to the formation of categories and the family of categories in the possibility space, fail to acknowledge the notion that language with words, connectors, grammar and syntax is based on conventions through evolution of intuitions and decision-choice needs of the social formation and production and reproduction of life. There is nothing naturally exact about the intuitive process in the space of acquaintance for the construction of the possibility space. In fact, what are epistemologically natural are inexactness, vagueness and ambiguities in the sending-receiving-interpretational module.

In fact, the whole of the linguistic process and the intuitive formation involve anticipations that connect the space of the universal object set to the possibility space is vague, inexact and ambiguous requiring different laws of thought from those of Aristotelian in their constructs and understanding. Intuition in conscious or unconscious or both states, does not allow the principle of *true or false but not both* [R14.97], [R14.100], [R19.3], [R19.4]. For intuition to be helpful in the knowledge construction process, it requires a balance between truth characteristics and characteristics of falsity where decision-choice actions of cognitive agents impose acceptance with attached *degree of intuitive confidence* which lies between zero and one. This degree of intuitive confidence through judgment is taken away by the Aristotelian *law of excluded middle* thus imposing a mechanistic personality of judgment. The Aristotelian laws of thought must be replaced with new laws of thought that are closer to the human linguistic character and reasoning behavior. These new laws of thought must present new reasoning as well as be able to incorporate the Aristotelian laws as extremes. The need for new laws of thought places on us to distinguish between *linguistic variables* and *symbolic variables* that include mathematical and logical symbols.

When the epistemic elements are moved to the possibility space by an epistemic process, they are transformed into *possibilistic epistemic elements* with associated degrees of anticipation for the next stage of cognitive processing. With the information on elements in the possibility space that is highly differentiated and relationally connected, cognitive agents see not $\{0,1\}$ but a distribution of degrees of complex relation between zero and one $[0,1]$ that helps to define shades of meaning, degrees of exactness that appeal to the fundamental analog, penumbral regions of interpretation and understanding, and others. It is here that the *fuzzy laws of thought* present an alternative for dealing with information processing and the knowledge construction at the presence of inexactness, vagueness and ambiguities. It is also here that the debate between the intuitionist logic and mathematic, on one hand, and the classical logic and mathematics based on the Aristotelian laws of thought, on the other hand, takes on an analytical importance for the knowledge production. The candidate for replacing the Aristotelian laws of thought is the fuzzy laws of thought which simply states that: *All propositions contain true and false characteristics in varying proportions where the acceptance of all true propositions and all false propositions is by decision-choice actions operating on a defined rationality* [R2.9, p. x]. The fuzzy laws of thought may be extended from the linguistic space to mathematical space, in that every element in the space of acquaintance belongs to a category with subjectively defined degree of belonging on the basis of quality and quantity. The fuzzy laws of thought incorporate the classical laws of thought as extreme cases. The fuzzy laws of thought are based not on dualism but on duality where the Aristotelian law of excluded middle is replaced by the law of *fuzzy continuum* [R14.5, pp.470-502].

Both the classical and fuzzy laws of thought involve two principles: one principle is used to construct significant categories while the other principle is used to determine which category may be taken as logically complementary to or follow the significant categories. The significant categories may come to be identified as the set of primary categories while the complementary categories are taken as derived categories from the primary categories. The laws of thought in category formation present us with two important variables in human thinking and manipulation for ideas and thought formation. They are linguistic or fuzzy variables and mathematical or logical variables. Thought systems require the presence of these variables and rules of combination in order to create significant propositions. Few distinguishing definitions of the different variables are required at this point of discussion.

**Definition 2.3.1: Classical Mathematical or Logical Variable**

The classical mathematical or logical variable in information representations, idea formations, and thought processes is one that moves over quantitative entities with fixed quality and meaning but not over degrees of meaning of variables or shades of truth, and is operated with exact laws of thought in reasoning. Its meaning is exact, its reasoning is crisp and its interpretation is unambiguous and free from vagueness, ambiguities varying quality and subjectivity.

**Note 2.3.1**

The classical mathematical and logical variables define every word or symbol as a point on the real number line or as a singleton set with one meaning and fixed quality (see the problem of Russell [R19.58],[R14.96] [R8.62], response by Max Black [R19.4][R19.5], intuitionist approaches and reflections by Brouwer, [R14.14], Dummett, R14.31] and Gödel's reflection [ R14.5, pp.447-469][R14.5, pp41-65]). In this respect, red as a symbol is exact in meaning and understanding. Thus, $x$ as a mathematical or logical symbol is no different from Red as a color defined in a spectrum. The classical variables are appropriate for exact quantitative analysis where quality is given and fixed. Varying qualitative dispositions with quantitative analyses are difficult for representation and reasoning. When time and dynamics are introduced, we mostly deal with *quantitative motion* without much to say about *qualitative motion* in terms of transformations. In other words, our classical mathematical reasoning has very little to provide us, if any, when qualitative-quantitative transformations are the focus of analysis. By the nature of exact classical reasoning, human action is externalized from the knowledge-production process and artificially grafted to human decision-choice action by the use of rigid rules contrary to linguistic reasoning and subjective judgment. The problem of classical mathematics in relation to qualitative motion was the concern of Karl Niebyl [R14.82] while the problem of qualitative disposition in inter-categorial conversions in social transformations was the concern of Kwame Nkrumah [R14.85].

To present a variable that incorporates qualitative and quantitative elements in its representation, we introduce the concepts of *linguistic variable* and *fuzzy variable* in representation and reasoning (for discussions on linguistic and fuzzy variables see [R3.52][R3.53][R4.34][R4.48] [R6.36] [R14.91] [R19.4]).

**Definition 2.3.2: Linguistic Variable**

A linguistic variable in idea formations, information representations and thought processes is one that moves over two elements of quantity and quality of entities with corresponding distribution of degrees of meaning or shades of truth or degrees of quality associated with the variable, with subjectivity of understanding and designed laws of thought that involve quality and quantity. Its meaning is inexact, its reasoning is vague and its interpretations are approximations with degrees of acceptance determined by decision-choice actions for the accepted level of exactness from the distribution of the degrees of exactness. The variable has also come to be known as *fuzzy variable*, the corresponding laws of thought for its manipulation, as *fuzzy laws of thought* and the corresponding mathematics for analysis and computations as *fuzzy mathematics*.

**Note: 2.3.2**

The fuzzy or linguistic variable specifies every word or symbol as a set of meanings with more than one element of meaning or quality or value. For example, the word red is a set where each element in the red has an associated degree to which it is said to be red and the cutoffs are decision-choice determined. Thus, every word or symbol is a set containing quantities or qualities or both with varying degrees of belonging where such degrees of belonging define the embodied linguistic or symbolic clarity. Its essential distinction from the classical variable is not only that it is a set, but that it allows the variable to take on conceptual meaning and analytical significance on the basis of human decision-choice action and judgment as endogenous processes in the knowledge-production system.

The point of special emphasis is that words, symbols and language are arbitrary and cannot take unchallenging claims to exactness as the natural order of things. In fact, if words are exact there would be no need for dictionary, definitions and explications. They are exact because we say so by limiting the domain of their conceptual meanings, analytical representations and areas of application. The domain that defines the conditions of exactness is decision-choice imposed for the cutoffs such that the boundaries are fixed by human actions. Intuition reveals to us that the elements in the universal object set exist only as interconnected elements and in continual transformation such that an object can manifest some characteristics in certain respect and conditions in such a way that the intuitive encounters may reveal differential attributes that accord them with some *epistemic identity*. In reference to the laws of thought in the construction of the possibility space (composed of families of categories), the classical laws of thought impose exactness of categories while the fuzzy laws of thought impose inexactness where the elements in the categories are decision-choice determined by subjective intuition with an associated degree of acceptance into a category. Another way of looking at the problem is that the determination of the conditions of exactness is internalized as part of the calculus of knowing through the fuzzy process.

The fuzzy laws of thought integrate human action into the process of category formation. They also reflect human intuitive limitations. Perfect exactness in all areas of human knowledge is cognitively unattainable. Perfect exactness is the ultimate goal that logic propels us towards. In this respect, it is useful to view exactness in degrees that are specified by a fuzzy set where complete inexactness and complete exactness are seen as polar cases of living duality in a continuum under logical tension. The possibility space for investigating natural phenomena in the epistemological space is exact to some degree that is determined in the knowing process through decision-choice action on the part of cognitive agents. This statement also holds for social sciences and other areas of knowledge search. The essential thrust in these discussions is that *unity of science* is a social attitude. Thinking under conditions of vagueness is an integral part of human thought process.

The problems of unity and separations in science are social. The solutions to these problems constitute important challenges of our knowledge enterprise in building the knowledge house with epistemic pillars. These epistemic pillars are constantly being strengthened for stability. The knowledge house contains many actual, potential and interconnected rooms whose numbers are constantly being increased, and whose room sizes are also under constant expansion with interconnected epistemic pipes to establish inter-sub-knowledge flow conditions. In this respect, the fuzzy paradigm offers us a new path to conceptualize the unity of science as well as provide us with a linkage path to intuitionist logic and mathematics, where continuity is a characteristic of the ontological space while discreteness is a characteristic of the epistemological space where discreteness is a creation by cognitive agents as a tool in the knowing process.

## 2.4 The Possibility Space and the Possible Worlds in the Knowledge-Production Process

Before we leave the possibility space and deal with the probability space, we will reflect on the contemporary conceptual scheme and logical development of the space of possible worlds and show their similarities with or differences from the possibility space [R12] [R12.2] [12.6] [R12.18] [R12.25]. The concept of the possible words may be viewed as a semantic characterization on a phenomenon and its existence or non-existence. Here, we have the realists who accept the existence of possible worlds and the antirealists who do not accept the existence of many possible worlds. The modalities of the possible worlds' logic towards knowledge construction involve possibility, necessity, contingency and impossibility which also appear in varieties as shown in Figure 2.4.1.

Given the modalities of possible worlds and varieties of the modalities, the collection of all possible worlds constitutes a set, $\mathbb{P}$, which we shall refer to as the *possible world set*. A number of questions arise in connection with the possible world set. What is a possible world? Is the possible world set finite or infinite in its conceptual construct? How the possible worlds are cognitively formed? To explicate the concept of the possible world set, we need to specify the defining characteristics of the constituent elements of the possible world set. In other words, what are the candidates for the possible world set and how do we select them? Similarly, how do we distinguish one element of a possible world from other elements in the set? Is each element in the possible world set defined by some empirical or axiomatic conditions? In other words, are the elements in the possible world set epistemic elements or transformational elements? For extensive discussions and debates on the possible worlds and the knowledge system, see [R12] [R12.9] [R12.10] [R12.11] [R12.16] [R12.27] [R12.28] [R12]. Are they formed by acquaintance and epistemic anticipations or by analytical extensions of propositions of what there is, or, the possible or the probable?

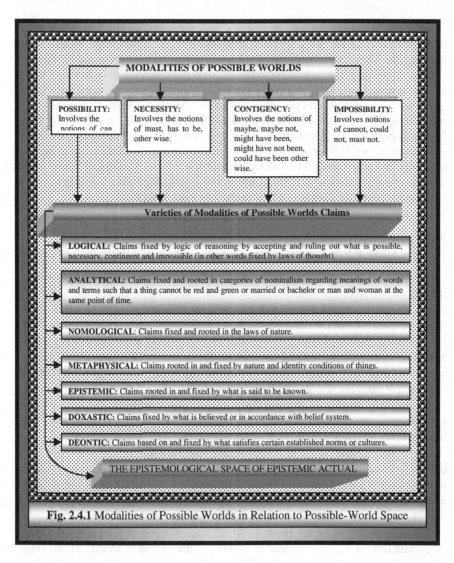

**Fig. 2.4.1** Modalities of Possible Worlds in Relation to Possible-World Space

In answering the questions raised, it is useful to notice that the possible world set, just like the possibility set, is conceivable in the epistemological space. The elements, however, may be related to some elements in the ontological space by reasoning through reductionism. In this process, attempts have been made to distinguish between possibility and conceivability by ruling out psychologism. Such a distinction is problematic in the sense that the justification of psychologism as it is related to the *laws of thought* has a defense in the basic foundations of reasoning and logic as conceived in the linguistic or symbolic manipulation with defined rules. The rules are established by epistemic conditions in order to arrive at a conclusion for a truth-value. No logic or defined set of rules of thought is independent of human cognition, and there is no logic that is not a provision of laws of reason or thinking, and hence, independent of human existence and decision-choice action of truth-falsity acceptance. From the viewpoint of human knowledge production, possibility cannot be separated from intuition, imagination and vision. But intuition, imagination and vision reside in conceivability. The separability and non-separability between possibility and conceivability will depend on the definition and explication of possibility and how it is logically constructed. Conceivability and possibility belong to quality-quantity continuum with neutrality of time.

When one accepts the existence of the universal object set, $\Omega$ or the cognitive potential space, $\mathfrak{U}$ , then the possibility and conceivability do not define the potential space. If one assumes possibility to be cognitively derived from the space of the universal, then one cannot dismiss conceivability and intuition from the construction of the possibility space. A clear distinction must be established as to the space in which our concepts apply. If acquaintance is required for conceivability, and conceivability is part of intuition, and intuition is a vehicle to establish cognitive possibility, then how do we separate conceivability from the possible world and the construct of the possible world set? In fact, the possible world set is spun by the conceivability process and the circumference of the possible world set, viewed as a circle, is determined by the diameter of conceivability. It is a dynamic and expanding set. Its foundation is affected by the state of what is taken to be known (that is, the available stock of knowledge) and defined in a continuum.

There are a number of problems in the possible world literature. It is difficult to assess the definition of possibility and possible world as they are used and developed. It is also difficult to place the concepts in the evolution of knowledge conclusion. What is the phenomenon of possible world and what is possible world theory? The modalities and types of modalities are arbitrary imposed without knowing what is assumed and what is taken to be logically derived from the universal system's duality of ignorance and non-ignorance. Is the possible world composed of propositions of possibility of differential existence as cognitively conceived, and on the basis of what? In relation to these discussions, we shall refer to definitions (2.1.1.and 2.1.2) as working definitions of the concepts of possibility and impossibility as elements of duality in a continuum and in relation to the concept of possibility space. From these definitions we shall examine how the elements in the possibility space may be related to the elements in the possible world set with different interpretations.

As defined, a possibility is a cognitive concept capturing anticipations over the path from the unknown to an epistemic actual. While the universal object set is independent of cognitive agents, the possibility space is dependent on the existence of cognitive agents. The concept being developed here, regarding possibility space, is not a collection of possibilistic propositions and cannot be subjected to propositional calculus. It is not semantic and cannot be subjected to definitional clarity. In other words, the possibility space is not the same as the possible world set on the basis of propositions. The possibility space is a collection of categories of elements and a family of families of categories that may be cognitively knowable. It may then include elements in the possible world set that may go through a knowledge test. The representational structures have been discussed in [R2.9].

It may be emphasized that in the framework of fuzzy paradigm and fuzzy laws of thought, every linguistic term as either a word or symbol is a set of degrees of meaning or value representations, but not a clear cut one-to-one correspondence to that which is defined and that which defines. This situation will violate our linguistic concept of synonym and antonym. This is one of the problem representations that Russell unsuccessfully sought to solve [R9.2], [R9.15],[R9.25],[R9.29] [R14.5] [R14.6] [R14.95], [R14.97], [R14.113], [R19.5], [R19.15] [R19.58]. To put it another way, every term or symbol in a formal or an informal language is vague. The exactness of a meaning of a term or symbol is obtained by explication through decision-choice action by cognitive agents except those designated as linguistic primitives [R9.20] [R9.21].

The point of concern is that theories of mathematics and rules of mathematical operations cannot be unquestionably claimed to be concerned with the structural properties of symbols independent of their meanings and the world of ideas, as sometimes claimed, especially by the mathematical formalists as opposed to the mathematical intuitionists. Similarly, mathematics and symbolic logic may be viewed as cognitive structures with the help of acceptably defined rules of operation and hence, they are branches of general grammar that allows successive categories to be derived and for meanings to be attached when the primary category is established. In this respect the rules of combinations of mathematical symbols are no different from the rules of combination of linguistic symbols. When mathematics is seen as contributing to the global knowledge structure, then within the theory of the knowledge square that is being presented here, mathematics must be seen not as knowledge, but as a way of organizing thought, just as any language for organizing thought about phenomena in their general aspects for deeper understanding of the phenomena through the analysis of their properties. Mathematics, just like symbolic logic, is an instrument of reasoning and it cannot be more than this. As an instrument of reasoning, it is not unique and may take many forms such as the form of the formalists or the form of the intuitionists or the form of fuzzy mathematics with corresponding grammars.

Similarly, propositions composed of inexact words or words with multiplicity of interpretive meanings cannot claim exactness in the true-false values. In so far as formalist's mathematical and logical symbols remain symbols under rule manipulations without reference to ideas and concepts, the claim of exactness in their derived conclusions can be assured. Such exactness cannot be maintained without some questions in the interactive system of the ontological and epistemic processes where cognitive agents are internalized as part of the knowing system irrespective of the tools of the knowledge production. In the physical systems, the cognitive agent places himself or herself as the subject of the knowledge search where the object under epistemic action is seen as independent from the cognitive agent. In the social systems, however, the cognitive agent is both the subject and object of knowledge search in complex interactive mode. In this way, the knowledge search process is divided into two. The former is an epistemic action on the knowledge about the environment of the cognitive agents. The latter is an epistemic action on the knowledge about the organizational arrangements of members in relation to the social and natural environments which are inseparably connected to the former in unity.

In this respect, mathematics is not knowledge about the elements in the universal object set and it cannot be so. It is knowledge to the extent to which it reveals paths for problem statement, analysis and computations, that is, it is a way to organize thought around the available information in the process of finding knowledge. However, mathematics can help and guide us to uncover aspects of knowledge about phenomena that we can collectively agree upon in the ontological and epistemological search. The classical mathematical and logical symbolism in cognition neglects human judgment throughout the knowledge production process. The problem that is faced in the knowledge production process involves the question: can mathematical and logical symbols be created that can avoid the qualitative structures, the vagueness in representation, and subjectivity in reasoning and ambiguities in interpretations and be epistemologically relevant and significant in all areas of knowledge production? If not , then what kind of representation of representation of information will be more helpful in organizing thought that makes allowance for human judgment under vagueness, ambiguities and approximation? The problem was picked up and discussed by Max Black in his treatment on vagueness [R19.4]. It is this problem of neglect by the formalists that formed the central discussions between the formalist and the intuitionist mathematics and logic and between [R14.5], Brouwer and [R14.14] Russell [R14.99]. It is also this problem of neglect that the fuzzy paradigm seeks to solve.

## 2.4.1  Classical Paradigm, Fuzzy Paradigm, Possibility Space and Possible Worlds

In this respect, let us consider the relative nature of the classical paradigm and the fuzzy paradigm with their corresponding mathematics and logics in organizing thought. The goal and objective of the classical paradigm with its logic and mathematics are no different from the goal and objective of contemporary development of fuzzy paradigm with its logic and corresponding mathematics. Both of them are ways of organizing thought, processing information and converting epistemic items into justified knowledge items with an associated degree of exactness. The classical paradigm accepts and operates on the principle of exactness and objectivity thus *externalizing* the subjective judgments of the cognitive agents from the knowledge-production process, where the cognitive agents merely follow an established set of deductive and inductive rules that leads to exact conclusions with a perfect degree of acceptance or non-acceptance. Such exact degrees of acceptance or non-acceptance may be assigned a value of either one (true) or zero (false) in the set $\{1,0\}$. In this way, the cognitive agent is the subject of the knowledge-production process, where all epistemic elements, excluding the knowledge seekers, are objects under cognition.

The fuzzy paradigm, on the other hand, accepts and operates on the principle of inexactness and subjectivity with vagueness, ambiguity and approximations where exactness and objectivity are cognitive agent's decision-choice determined thus *internalizing* the cognitive agent as an integral part of the knowledge-production process. In this way, the cognitive agent is both the subject and object of the knowledge-production process. The degrees of acceptance of true-false propositions are less perfect and require qualifications of measures in degrees of values in a continuum scale between zero and one as extreme cases that are inclusively specified as an infinite set of the form, $[0,1]$, and where such measures include subjective judgment through decision-choice actions at all stages of the knowledge-production process. The fuzzy paradigm accepts the universe in the fundamental continuum of the ontological space and then designs methods and techniques that persevere this continuum in the epistemological space where methodological discreteness is viewed as approximations of the fundamental continuum, and methodological continuum is viewed as an enveloping of the methodological discreteness. The classical paradigm designs methods of discrete processes to study the continuum phenomenon. Here, it is not clear whether the ontological space is viewed as fundamentally discrete where its study must follow methodological discreteness.

If science is defined and demarcated with the properties of classical mathematics as exact representation, then our knowledge production will be in trouble where there is an open door for any knowledge researcher to find a mathematical concept and its algorithm and search for or make up a problem that fits it. Here, one must keep in mind the intense discussions on the possible world and how such discussions may relate to mathematical objects and reasoning. A conclusion may, therefore, be derived that is mathematically true within its logical system, but have little to say about social or natural phenomena. Such a situation has happened many times before and will continue to happen in economics and other areas of social sciences where mathematical characterization is considered as elegant, abstract and intellectually appealing and contributing. It is here that some care must be exercised on what constitute mathematical truth and what constitute truths in natural and social sciences.

## 2.4.2  *Mathematics in the Classification of Science and the Epistemic Process*

In the use of unconstrained mathematicalization, mathematics drives the problems where unnecessary assumptions are made about the phenomena in order to fit the problems into mathematical structures rather than the problems driving the development of the needed logic and mathematics for abstracting solutions. The insistence on the conditions of exactness and objectivity will exclude a number of accepted areas of exact science and place them into inexact science. If science, however, is defined and demarcated with the properties of fuzzy mathematics and its algorithms as inexact representations with subjective judgment, then our knowledge production will cover wider areas of sciences. The strength of fuzzy paradigm in the knowledge-production process may be seen in two actions. It provides an allowance to deal with quality, vagueness and ambiguities in thought and cognition. It also provides an allowance in dealing with classifications of degrees of exactness in acceptances of conclusions due to the processing of *defective information structure* through subjective decision-choice actions in the deductive and inductive processes of thought. We shall define, explicate and expand on the nature of defective information structure and how it affects the use of mathematics in theoretical and applied knowledge (one can also refer to [R11.9]]).

Mathematics and logic are epistemic tools that allow simpler symbolic representations of complex concepts, ideas and expressions of linguistic nature for easier manipulations under the guidance of their own grammars that are not possible by the use of the grammar of ordinary language. Mathematics, thus, offers an expanding set of rules of combination and logical manipulations of symbolic representation of concepts, ideas and further derived system of ideas that help to reveal the differences and unity of all sciences. Easier manipulations of concepts, thought formations and rules of combination of symbolic representations must not be equated with exactness of knowing. Mathematics and logic are used to organize the elements in the possibility space for further analysis in other spaces in the knowledge-production process. In this respect, it may be argued that every homological, metaphysical and analytical element in the possible world set is logically reducible to an element in the possibility space through a process. Alternatively stated, the elements in the possible world set are logically derived categories where their primary categories are traceable to the possibility space. For mathematics to fulfilled the task of unity of science and be applicable to all areas of sciences, it must be able to represent all information in terms of defective and non-defective structures.

The possible world set is obtained by combinatorial construction from the epistemic elements in the possibility space where the set of its primary category may be obtained by methods of reductionism from the elements in the *possible world space* that houses the possible-world set. The possibility space constitutes the family of primary categories while the possible world set constitutes the family of derived categories from the possibility space. The set of the primary categories of the possible world set is obtained by methods of reductionism while the set of derived categories in the possible world space is obtained by the methods of constructionism. The most important distinguishing factor between the possibility space and the possible world set or the possible world space is that the elements in the possibility space involve processes for verifications, falsifications and corroborations of knowledge and truth irrespective of how the elements are abstracted, define and analyzed from the universal object set. The elements in the possible world set must satisfy the verification, falsification and corroboration conditions of logical truth but not necessarily the conditions of knowledge. Furthermore, any possibilistic epistemic element in the possibility space becomes a set of possible world elements that are generated by imaginational extension into the possible world space which then constitutes a family of derived categories. The size of each derived set for any possibilistic epistemic element, therefore, depends on imagination.

It may be noted that the elements of the possible world set are semantic in nature where the logical truths of their elements are the subject of analysis. Their knowledge contents are not the subjects of analyses in the possible-world space. The size of the possible-world space is also amplified by cognitive imagination from the elements of the space of the epistemic actual. An element in the possible world set may satisfy conditions of either logical truth or falsity but may have nothing to materially contribute to the state of knowledge. It may exist as an imaginative element in correspondence to possibilistic epistemic element. For any element in the possibility space and any element in the space of the epistemic actual, there are technically infinite propositional elements that may be generated by imagination into the possible world set that is a collection of possible worlds associated with them. The distinguishing characteristic between the possibility space and the possible world set may be viewed as the relationship of propositional truth to knowledge. A proposition may be true in a given axiomatic system and may have nothing to do with knowledge. It is simply an exercise in reasoning that may or may not have any knowledge content as we observe in abstract mathematical and logical systems.

Mathematics and corresponding theories just as symbolic logic and corresponding theories may equip us with sophisticated computability and reasoning systems, but may have nothing to tell us about conclusions of knowledge verification, falsification and corroboration as we move from the universal object set to what is claimed to be knowledge items. They may assist in designing computable test principles when concepts are placed behind the symbols in use. The results of pure mathematics and logic are not knowledge (in ontological sense from the epistemological space). Their study is to equip the holder with a discipline in the use of rules of reasoning as well as their further abstract development when such rules are constructed. They help us in abstracting significant propositions about elements in the possibility space. They also help in defining the structural unity of all sciences as well as identify some common characteristics and differences at the most diverse rooms of the general knowledge house. They, however, say nothing about the knowledge contents of the propositions about elements in both the possibility space and the possible world set or the possible-world space. The possibility space $\mathbb{P}$ is contained in the possible world set $\mathbb{P}$ where $(\mathbb{P} \subseteq \mathbb{P})$. The size of $\mathbb{P}$ in $\mathbb{P}$ will depend on the imagination of cognitive agents. The possible world set may be enlarged by increasing the number of semantic propositions about a primary element in the possibility space.

## 2.5   The Possibility Space, the Primary Category and the Derived Category of Knowledge

The possibility space has been presented as the second building block in the chain of the process of knowing. It is connected to the universal object set through the relationship between the perception and reality supported by the cognitive process. The *universal reality* is that which is there as distinct from *epistemic (cognitive) reality* of that which is claimed to be known. The universal reality is also ontological reality. The universal reality is the reflection of the *ontological space* while the epistemic reality is the reflection of the *epistemological space* through the cognitive process. The knowledge-production process begins with the universal reality and ends with the epistemic reality which is connected to the universal reality by logical transformations of categories of epistemic elements. Here, as elsewhere, we must separate *natural categories* from *logical categories*. Perception and the universal reality on one hand, and universal reality and epistemic potential on the other hand constitute categories of polarity and duality. The epistemic potential and the logical process, in turn, constitute a change in the knowledge structure of the *universal unity* through epistemic conflicts of rational justification, corroboration and verification with their corresponding appropriate index measures in the polarities and dualities.

## 2.5.1 Essential Postulates of the Primary and Derived Categories

The universal reality is partitioned into categories of unit existence that allow different forms of elements to be identified, distinguished, classified and named on the basis of characteristic sets of elements which form the nominal basis of language. Given the family of categories as defining the starting point of a search for knowledge about the universe, we work with conditions of epistemic initialization of a category called the *primary category* in the family. We must keep in mind that our awareness in the universal planetary system is the work of information, epistemic process and decision-choice actions.

### Postulate 2.5.1: Existence of Primary Category

In the process of knowing, the set of categories of the possibility space has a *primary category* from which all other categories of epistemic reality evolve themselves through logical substitution-transformations into sequences of *derived categories* that must satisfy the principles of justification, corroboration, verification and falsification.

### Postulate 2.5.2: Non-uniqueness of the Primary Category

The primary category of the possibility space is not unique in the logical transformations in the sequence of derived categories of epistemic reality. Every derived category has a supporting primary category and every element in the possible world set has its supporting primary category from the possibility space. Every derived category may constitute a primary category for subsequent derived categories in the sequence of the knowledge production.

These postulates merely present to us the idea that any element of the possibility space is a candidate for selection by cognitive agents as a primary category of logical operations for truth and knowledge justification, corroboration and falsification as we work ourselves toward epistemic reality. The decision-choice action on the primary category will be influenced by what is believed, known and accepted in the social structure. The collection of all candidates of the primary categories constitutes the possibility space in the knowledge production process. The elements in the possibility space are called the *possibilistic epistemic elements.* An example of the choice of a primary category is Thales' choice of water as constituting the foundation from which all things are derivable. To Thales, therefore, water is a primary category from which all derived categories must meet justification, verification and corroboration principles for knowledge acceptance. Similarly, for Berkeley, the primary category was spirit and that every element in the possibility space is seen in terms of a spirit, [R14.85]. It is possible to select energy as the primary category of knowing. Basically, the choice of a primary category for research in the knowledge-production process is to answer the question of "what there is". The justification principle in the selection of the primary category may be empirical or axiomatic, depending on the nature of the subject area of knowledge search. In physical sciences, it is primarily experimental and empirical; in economics and decision sciences, it is primarily axiomatic and mostly non-experimental.

### 2.5.2   The Establishing Characteristics of the Primary and Derived Categories

The characteristics of the primary category in the possibility space may be established on the basis of either empirical principle or axiomatic principle or both from the available information structure. In each case, we encounter conditions of inexactness in representation and vagueness in interpretation of exact structure of the primary category through its defining characteristics (attributes). The conditions of the primary category are about ontology that must be related to the epistemological process of explanation and prescription. It is within the elements in the possibility space that visions and imaginations acquire meaning in the *prescriptive science,* and understanding and reality acquire clarity in the *explanatory science* as they are related to the sequences of the knowledge production. The empirical or axiomatic conditions merely help to establish existence of a possibility from which its connection to the universal reality must be shown. The primary category is not knowledge and its claim to knowledge must be shown or demonstrated as belonging to the universal object set. The elements in the possibility space initialize the knowledge-production process.

The construction of the elements in the possibility space must be shown to be composed of elements with their corresponding degrees of possibilistic belief. Every element in the possibility space is a category and corresponding to it is a distribution of possibilistic uncertainty due to vagueness and limitations of ambiguity that are imposed by inexactness of acquaintance and measured by *possibility index.* In other words, the vagueness, ambiguities and inexactness are the result of the information-transmission and receiving process. The variable that tends to capture the categorial element is referred to as the *fuzzy variable.* The field of variation of a fuzzy variable is fixed by the distribution of the values of the possibility index that impose the boundaries of applicable areas of the subjective possibilistic belief. The boundaries may be made elastic with the expanding knowledge structure. The elements in the possibility space are then ranked in terms of the measures that capture the index of possibilistic belief. Acquaintances create primary information which may be called *experiential information structure* that defines the elements in the possibility space. Imagination creates derived information on the basis of the primary to give us another type of information which may be called *imaginative information structure* that characterizes the possible-world elements.

Out of the possibility ranking of the elements in the possibility space, a selection is made of the primary category that satisfies the highest index of the possibility belief. The possibilistic belief captures uncertainty in the initial knowledge production due to vagueness of meaning of symbolic representation such as words and ambiguity in propositions or numerical complexity, or what may be referred to as *inexact problems* of *linguistic variables* for approximate reasoning and communication. The possibility space is such that when one selects the primary category, relative to a phenomenon, then one has in mind the notion that all other categories related to the phenomenon in the universal object set can be shown to be derivatives from the primary category. These primary-derivative categorial relations of the elements in the possibility space require the specification conditions of *categorial conversion* in the possibility space [R2.9]. The concepts and the representations of ideas contained in the fuzzy variables allow the development of substitution-transformation dynamics as to how the knowledge production works through the decision-choice rationality under vagueness, ambiguities and volume limitation of information limitation, which may generally be represented as *defective information structure* [R11.19]. The exactness or inexactness of our knowledge is rooted in our need for better understanding of the concept of defective information structure; our lack of qualitatively and quantitatively good knowledge in any area of the knowledge system reflects our difficulty of internalizing and integrating the ignorance of cognitive agents as part of the knowledge-production process, either through methodological constructionism or methodological reductionism which must transform an information structure to a knowledge system. This information structure has two components of quality and quantity. The deficiency of any one or both produces defective information structure that must be subjected to an epistemic process for knowledge derivation. Let us provide a working definition of the defective information structure.

### Definition 2.5.2: Defective Information Structure

A defective information structure is one that contains qualitative characteristics of vagueness, ambiguities that give rise to fuzzy beliefs with possibilistic uncertainties and fuzzy risk, as well as quantitative characteristics of volume limitations that give rise to stochastic beliefs with probabilistic uncertainties and stochastic risk in the knowledge-production process.

## 2.6  Possibility Space, Explanatory and Prescriptive Sciences in the Knowledge-Production Process

To place the theory of the knowledge square in the general epistemology of the knowledge-production process and how it will help us to understand the empirical and axiomatic foundations of exact and inexact sciences, it is useful to partition the space of science into *explanatory* and *prescriptive sciences*. It is also useful to relate the possibility space to the analytical construct of explanatory and prescriptive sciences. In the explanatory science, the existence of an element, belonging to the possibility space, is examined or established by either empirical or axiomatic principle to postulate its ontology or knowability of *what there is*. Additionally, the characteristic behavior of *what there is* becomes subjected to explanability. The explainable behavior becomes subjected to the conditions of predictability. Thus, the chain of the cognitive process proceeds from ontology to knowability to explanability and to predictability. In other words, ontology (that which is there) precedes epistemology (that which is claimed to be known). This is shown in Figure 2.6.1

We have pointed out that the possibilistic categories are established by means of the fuzzy paradigm where vagueness, inexactness and ambiguities due to *intuitive processes* (intuition) are essential characteristics of any possibility category. In terms of explanatory science, a fuzzy category comes to us with a degree of possibilistic uncertainty that is associated with the belief that the item is likely to be knowable. Each possibility category may be amplified into a possible world and placed into the possible-world space. In this respect, all elements in the possible world set carry with them defective information structures.

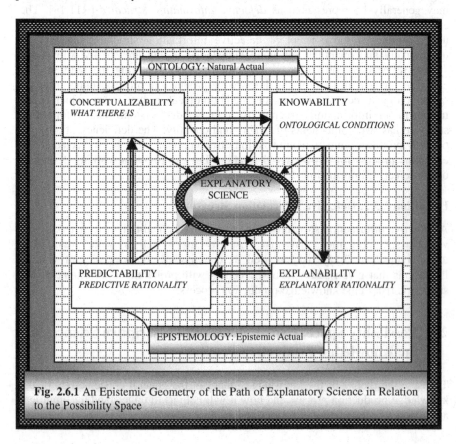

**Fig. 2.6.1** An Epistemic Geometry of the Path of Explanatory Science in Relation to the Possibility Space

The logical system that presents this epistemic structure of the process of knowing constitutes a foundation of the explanatory theory within the *explanatory science* [R14.35] [R14.50]. The cognitive approach to the explanatory reasoning and its analytical structure, in the process of knowing, constitute the *explanatory science*. In the explanatory science, the decision-choice process is such that the development of the primary category and the derived categories must assume the existence of *what there is* while the explanatory theory establishes an explanatory rationality in support of the justification of knowledge claims of *what there is*. We may refer to *what there is* as an *explanatory possibilistic category* that is to be subjected to the logical principle of knowing from the possibility space. The possibility space, in this case, is composed of elements of possible knowledge with possibilistic uncertainty. The elements do not have claims to knowledge except claims to knowledge possibilities. The cognitive task, at the level of ontology, is to construct a possibilistic belief system in relation to the possibilistic uncertainty as a justification of *what there is*, and at the level of epistemology, to construct a justification in support of possibilistic belief indexes as measures of degrees of knowability.

Different interpretations of the possibility space and the possibilistic categories present themselves when we enter into the realms of *prescriptive science* [R14.30], [R16.7] [R16.12]. In the prescriptive science, the decision-choice process is such that the development of the primary and derived categories must assume the existence of *what ought to be*, while the *prescriptive theory* establishes the conditions of the *prescriptive rationality* for the justification of *what ought to be* that meets the conditions of possibilistic belief in terms of candidates for actualization. The possibility space is composed of desirable elements that fit into human vision and realistic imagination for transformation into the space of the actual through a process. The possibility space is seen to be composed of elements not for understanding but to be transformed into elements in the category of an actual that must reside in the space of the actual. The conceptual integrity of any possible-world construct in the possible-world space enters here, where the conditions of actualizing an element in the possible world set may be established by the method of prescriptive theory, to the extent to which the element meets the conditions of categorial convertibility. Let us keep in mind the discussions of the similarity and differences between the possibility space and the possible-world space as seen by the definitional characteristics of their existence.

The elements in the possibility space do not have claims to epistemic actual except for simple claims into the transformation-substitution process under the logic of categorial conversion. The cognitive task of prescriptive science, at the level of ontology, is to construct a possibilistic belief system that will provide a justification of *that which ought to be*. At the level of epistemology, the cognitive task is to construct a justification in terms of possibilistic indexes for degrees of categorial convertibility, or constructability or reductionability that will ensure appropriate transformations of the possibilistic epistemic elements. Here, we speak of the conditions of categorial convertibility in the process of either construction or reduction.

It is important to keep in mind the interpretational framework of the possibility space and possibilitic categories as they relate to the explanatory and prescriptive sciences within the theory of the knowledge square and how exactness may be derived from inexactness. The elements of the possibility space are defined by fuzzy characteristics and categories. The fuzzy characteristics may be seen in terms of possibilistic belief systems with corresponding possibilistic categories and fixed by possibility distribution. Let us illustrate the deferential interpretations of the possibility space as they are related to explanatory and prescriptive sciences. Imagination, intuition, vagueness and ambiguities are subjectively transformed into decision-choice actions as internal parts of the knowledge-production process.

Consider an element called good society in the possibility space. A good society may be taken as the primary category in the possibility space from which different categories of good society may be derived. At the level of explanatory science, a good society may be considered first as *what there is* for knowability in the framework of ontology as established by empirical or axiomatic conditions. Is there anything called a good society? This question has a follow-up one. How many good societies are there in the possibility space and how similar are they? It is, therefore, a possible element for cognition. Given the existence of a good society, we must construct an explanatory theory of its behavior and then predict its further behavior in the same environment or its actualization in a similarly created environment. In the explanatory science, the good society is empirically established, and its predictability must take place in the same environment or another environment with identical structure. At the level of prescriptive science, the good society is by a value judgment existing in the possibility space or in the possible-world space where its characteristics are established by axiomatic system that may be supported by a set of sense data on the basis of a justified possibilistic belief. The element a *good society* may belong to the possible world set. As in the explanatory science, a *good society* is accorded its primary category from which derived categories are also made in the prescriptive science.

In the prescriptive science, the task of cognitive agent is not to establish *what there is*, explain and predict its behavior. The task is to establish *what ought to be* by subjective assessment of the conditions of *what there is*, and how to transform it into the potential and actualize *what ought to be* in the actual-potential duality. The bringing into being of what ought to be requires the provision of a prescriptive theory that defines a path or a set of conditions to transform *what ought to be* into *what there is. What there is,* is a knowable possibility and *what ought to be* is an actualizable possibility. In both cases of explanatory and prescriptive sciences, the set of categories of possibility, in relation to a particular phenomenon, is closed by decision-choice action to constitute a manageable possibility set. Generally, however, the possibility set is open. The general possibility space is convexly open and it is this convexity property that allows possible worlds set to be formed in the sense that any convex combination of derived categories from either the possibility space or the space of the epistemic actual belongs to the possible world set. For example, a living dog with one head belongs to the space of epistemic actual while a living dog with an elephant head or with multiple heads belongs to the possible world space. The possibility set is constructed from the possibility space and it is closed by decision-choice action where its elements are examined for probable categorial conversion through a substitution-transformation process. The possibility set is a categorial member of the possibility space and it is finite relative to a phenomenon. The possibility space that indicates the possible knowable elements, at any moment of time, is also finite, where such finitude is decision-choice determined, while the possible-world set is infinite and only constrained by imagination and vision.

At the level of explanatory science, this substitution-transformation process takes place through theoretical constructs to provide a justified belief for the claim of *what there is*. The behavior of what there is and its further occurrences under fuzzy information constraint are contained in the possibility space. The explanatory claims appear as propositions and hypotheses. The acceptances of their validity must be examined against the *space of the probable* for the phenomenon in question. At the level of prescriptive science, however, the substitution-transformation process takes place also through theoretical constructs to provide a justified belief for the claims of *what ought to be* and its actualization under a fuzzy information constraint which characterizes each element in the closed possibility set regarding a particular phenomena. The possibility set for each phenomenon is closed by fuzzy decision-choice rationality with the principle of fuzzy decomposition; otherwise, it is infinitely open. The chain of the cognitive process in the prescriptive science proceeds from conditions of *conceptualizability* to *prescriptability* to *implementability* and then to *actualizability*. In other words, epistemology (that which is claimed to be known) precedes ontology (that which is there). It is here that the possible world set enters the knowledge-production process. It is also here that engineering of all forms enters into the knowledge production enterprise. An immediate question follows as to whether the results of engineering fall under epistemology or ontology? The path of prescriptive science as seen from the possibility space is illustrated in Figure 2.6.2 In relation to epistemology and ontology.

The possibility of the knowable in thought arises from the nature of intuitive form that reflects the first level of cognitive encounter through acquaintance with some elements in the universal object set. The role of intuition as presented in the formation of categories in the possibility space should not be confused with the intuitionist mathematics and logic even though they have something to contribute in strengthening the conceptual representation and understanding of the role of intuition in cognition and symbolic reasoning.

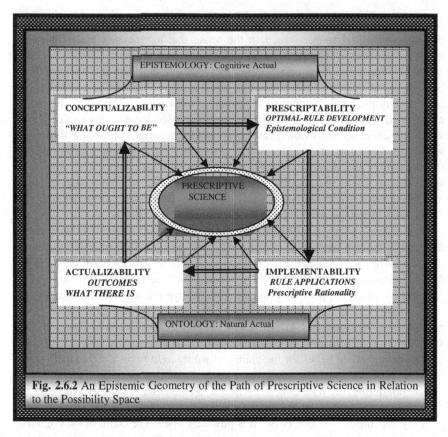

**Fig. 2.6.2** An Epistemic Geometry of the Path of Prescriptive Science in Relation to the Possibility Space

The epistemic geometries of the paths of explanatory and prescriptive sciences in relation to the possibility space have interesting cognitive distinction. In the explanatory science, we begin from the *potential space* of knowing, locate the possible element of what *there is*, and move it to the possibility space for knowing. Ontological conditions that present a belief of knowability are constructed. We then develop explanatory theories about the phenomenon of interest and subject them to conditions of prediction, falsification, verification and corroboration [R2.9][R14.11] [R14.13][R14.21] [R14.30]. We simply move from the potential to the actual through an explanatory process. The cognitive process of moving from the potential to the epistemic actual by explanation constitutes the *explanatory rationality*. There is no cognitive effect on *what there is* except the construction of the justification of its existence and understanding of its behavior after its existence is cognitively ensured. In the explanatory science and the corresponding theories, the focus is explanation and understanding of the behavior of the elements of what there is but not to alter them or improve what there is.

In the prescriptive science, we also begin from the space of the epistemic actual, locate *what ought to be*, and move it into the possibility space through the possible world set, construct prescriptive conditions, in terms of prescriptive theory, that present a justified belief of its happening, in terms of becoming *what there is*. We then develop strategies for the theory's implementation and act on the conditions for the actualization of *what ought to be*. The cognitive process is referred to as *prescriptive rationality* and the actualization of *what ought to be* becomes *what there is*. In the prescriptive science, therefore, there is a cognitive effect on *what there is* in that *what ought to be* is cognitively acted on and transformed to become *what there is* as an ontological element.

The test of validity of prescriptive theory is the success in transforming *what ought to be* (the potential) into *what there is* (the actual). The job of prescriptive science is to change the constituent elements of the space of the actual and link it to the potential space. The prescriptive theory becomes an explanatory theory when the implementation is successful. The job of explanatory science is not to change the elements in the space of the actual but to provide a justification and explanation that link the space of the actual to the space of the potential. The explanatory theory may provide some prescriptive conditions for the actualization of the same phenomenon in a similar environment [R2.9] [R14.30]. We shall now turn our attention to enter into the conceptual zone of the probable in relation to the epistemic path of the knowledge square.

# Chapter 3
# The Theory of the Knowledge Square and the Concepts of Probability and the Probability Space ( $\mathfrak{P}$ )

We have discussed the potential and possibility spaces and their relationships to the knowledge-production process as viewed from the theory of the knowledge square. These conceptual relationships were connected to the possible-world space, the explanatory science, prescriptive science and the defective knowledge structure. On the basis of the defective knowledge structure, the roles of explanatory theory and the corresponding explanatory rationality were projected. Similarly, the prescriptive theory and the corresponding prescriptive rationality were discussed .We shall now turn our attention to the probability space and show how it is connected to the possibility space and then to the universal object set in a backward regressive process of the knowledge-production process. The possibility space and its construct on the basis of the fuzzy paradigm to deal with the problem of possibilistic uncertainty provide us with the analytical structure of possibilistic reasoning under conditions of defective information structure that is constrained by quality and quantity deficiencies about epistemic elements with neutrality of time.

## 3.1 The Knowledge Square and the Concept of the Probability Space $\mathfrak{P}$

Much of our knowledge production has benefited from *probabilistic reasoning* and logic that follow the classical laws of thought. The essential point that we must note is that *possibilistic reasoning* and its logic precede probabilistic reasoning and its logic. The possibilistic reasoning is always assumed to exist in the same way that we assume the existence of the possibility space on the basis of which the probabilistic reasoning is developed. In our mathematical constructs, probability and probabilistic reasoning that are associated with conclusions in exact sciences, we must assume the existence of the possibilistic set that is developed from the possibility space for a given phenomenon. The implication is that the possibility space with its elements is defined and the possibility set for any

K.K. Dompere: The Theory of the Knowledge Square, STUDFUZZ 289, pp. 43–54.
springerlink.com                      © Springer-Verlag Berlin Heidelberg 2013

phenomenon is given and closed for the construct of the probabilistic reasoning. It is on this basis that impossible outcomes and sure outcomes, relative to a phenomenon, take on scientific meanings in the probabilistic reasoning (see alternative discussions in [R12.6]).

It may be noted that without the possibility space, possibility set and possibilistic reasoning, a search for knowledge or scientific reasoning may have no meaning and social value in thought and the practice of thought. A cure for cancer is a phenomenon and this phenomenon is defined in the possibility space, and various techniques and methods to reach the cure constitute derivatives with distribution of possibilistic measures of attached beliefs in the possibility space. The cure for cancer is a potential that we have moved by some method of cognition into the possibility space that provides us with a possibilistic belief system for epistemic investigation to find it.

## 3.2   The Concept of Probability and the Construction of the Probability Space

From the possibility space and given the primary category, a set of derived categories is abstracted by the method of categorial conversion and then examined for inclusion into a possibility set for any given phenomenon. Each element in a derived category $\mathbb{C}$ of the selected phenomenon, comes with a measure of possibilistic belief, $\pi$, which may be referred to as *possibility measure*, $\mu_{\mathbb{C}}(\pi)$ that depends on the possibilistic belief given the cognitive transformation from the universal object set. It may be noted that the concept of belief is a linguistic variable that has a fuzzy covering in its measure of degrees of belonging to a category. The possibility set, relevant to an event or a phenomenon, broadly speaking, is then constructed by imposing a cutoff by either a method of fuzzy ranking or optimization through the membership characteristic functions that may be constructed by subjective or objective conditions or both. The logical process of constructing the closed possibility set for each phenomenon works through the principles of duality, opposites and continuum. The opposites are the extremes of possibility and impossibility. Each element is conceptually viewed as a set of possibilities with a distribution defined by membership characteristic function. The same element is also viewed as a set of impossibilities with a distribution defined with a membership characteristic function. For the same element from the universal object set, as the size of the characteristic set of impossibility decreases, the size of the characteristic set of the possibility increases in dualistic sense.

In this way, the possibility and impossibility are viewed in terms of *duality* in the fuzzy thought system with the *principle of continuum*. Thus, for every degree of possibility, there is a corresponding degree of impossibility for every epistemic element in such a way that every degree of possibility has a support of a degree of impossibility and vice versa. Both of them reside in an epistemic unity without which their conceptual understandings are meaningless. They are, however, viewed in terms of *dualism* in the classical thought system with the *principle of excluded middle* where the element is either possible or impossible without an indication of how the possibility set is obtained. The possibility in the fuzzy system is established by a positive characteristic set with a corresponding membership characteristic function that is increasing in the characteristic set. Similarly, the impossibility is established by a negative characteristic set with the corresponding membership characteristic function that is decreasing in the characteristic set. The membership characteristic function of each category, belonging to the possibility set, is taken as the goal and the membership characteristic function of its impossibility set is taken as a constraint to create a fuzzy optimization decision problem. The fuzzy optimization decision problem is then solved to obtain an optimal-level value which is then taken as a fixed optimal indifference value. The closed possibility set is then constructed as a fixed-level set with the optimal value, and the method of fuzzy decomposition, for a given phenomenon. The optimal fuzzy membership values are then used to rank the categorial derivatives that are related to the phenomenon of interest. This set involves the concept of epistemic possibility. The method of the construct is the same for both the explanatory and prescriptive sciences in the possibility space. The meanings, differences and similarities of the concepts of duality, dualism, and their corresponding principles of continuum and excluded middle respectively are discussed in the various chapters in this monograph. Some definitions are needed to build the relational structure of probabilistic concepts to the possibilistic concepts.

### Definition 3.1.1: Closed Possibility Set

A possibility set of a phenomenon from the possibility space is said to be closed if there exists an optimally minimal membership value $\alpha^* \in [0,1]$ that is a support of the ordinary possibility set of the phenomenon, where the value $\alpha^* \in [0,1]$ is decision-choice determined and it is bounded from below by $\alpha^* < \alpha \in [0,1]$; otherwise it is said to be open.

### Note 3.1.1

Any open possibility set is linguistically a subset of the possible world set. The set $\alpha \in [\alpha^*, 1)$ is the support of the possibility set closure. Every probability set has a closed possibility set that allows the probabilistic reason to be valid between impossible and sure events. Every epistemic element with a possibility index-value $\alpha \in (0, \alpha^*)$ is taken as an impossible event with probability value equal to zero in the probability space. We must keep in mind that impossible events may become possible while possible events may become impossible as time proceeds and new qualitative and quantitative information characteristics become available to enhance the domain of the possibility space.

### Definition 3.1.2: Probability

Probability, in the conceptual scheme of the knowledge square is that which is cognitively abstracted from the possibility space through an epistemic transformation of the conceptual possibility, with justified degree of belief regarding the likelihood that the possibilistic epistemic element may be turned into a justified knowledge item. The improbability is that which has justified degree of belief that excludes it from the likelihood of transforming the possibilistic epistemic element into a knowledge item.

### Definition 3.1.3:  The Probability Space

The possibility space in the conceptual scheme of the theory of the knowledge square is a conceptual subsystem that provides us with a linkage between the possibility space and the space of the epistemic actual that allows the understanding of the knowledge transformation process in the epistemological space where the possibilistic epistemic elements are cognitively believed to be informationally probable as a knowledge item that can enter the space of the epistemic actual with a degree of *stochastic uncertainty*. The collection of all elements that are moved from the possibility space with stochastic justified belief constitutes the probability space with elements called *probabilistic epistemic elements* that may be specified as:

$$\mathcal{B} = \left\{ \left( y, \mu_{\mathcal{B}} \left( y \right) \right) \mid y \in \mathcal{P}, \text{ and } \mu_{\mathcal{B}} \left( y \right) \in \left[ 0,1 \right] \right\}$$

Where

$$\begin{cases} y = \left( x, \mu_{\mathcal{P}} \left( x \right) \right) \text{ is a possibilistic epistemic variable from the possibilisty space} \\ \mu_{\mathcal{P}} \left( x \right) = \text{the degree of possibility of knowledge, a measure of possibilistic belief} \\ \mu_{\mathcal{B}} \left( y \right) = \text{the degree of probability of knowledge, a measure of probabilistic belief} \end{cases}$$

Given the closed possibility set associated with a relevant phenomenon in the knowledge-production process, and with specified degrees of possibility constructed on the basis of characterized linguistic variable, an approximate reasoning and fuzzy paradigm, we now raise a question about the elements in the closed possibility set. The question is: how probable will any item in the possibility set be known from the explanatory rationality or be actualized from the prescriptive rationality? Let us notice that every epistemic item is an event with a complex combination of possibilistic and probabilistic uncertainties. These epistemic items may be restructured into what is called an event space. The search for an answer to this question points us to another epistemic uncertainty that is due to quantitative information limitation, in terms of volume *limitationality* and *limitativeness,* which surrounds the elements in the possibility set for a particular phenomenon. Let us keep in mind that the elements in the closed possibility set come as dual and as a fuzzy variable of representation, where each element appears with its degree of belonging, where such degrees of belonging capture the conditions of vagueness, ambiguities, inexactness and subjectivity through vision and imagination which together represent the *possibilistic uncertainty*. The possibility space comes with its own topology. It is, thus, a cognitive construct on the basis of a particular epistemic transformation to help eliminate some doubts of possibilistic uncertainty, and to decide on the degree of possibilistic uncertainty that is tolerable for an acceptance of an event as reasonably knowable under *explanatory system*, or reasonably actualizable under *prescriptive system* within some degree of human possibilistic ignorance associated with a particular phenomenon. In moving from the possibility space that is characterized with some degree of fuzziness, we must deal with an information limitationality and limitativeness in the global field of knowing in terms of volume that is available for quantitative processing. At this analytic point, there is a need, therefore, for the definitions of the concepts of limitationality and limitativeness as they relate to information structure.

### Definition 3.1.4: Information Limitationality and Limitativeness

A decision-choice system $\mathbb{D}$ in the knowledge-production process is said to be information *limitational* (*limitative*) if an expansion in the volume of the supporting information structure is a necessary (sufficient) condition in obtaining a critical level of knowledge acceptance.

### Note 3.1.2

The information limitationality and limitativeness, defined in terms of quantitative characteristics, present uncertainties which are different from those of possibilistic uncertainties that are due to the global fuzziness. It is called *probabilistic uncertainty* because it involves probabilistic or stochastic belief system due to incompleteness of the volume of information. To deal with probabilistic belief, an epistemic mapping is called upon to transform the closed possibility set into probabilistic categories. The collection of all these probabilistic categories constitutes the probability space that comes with its own topology. Each element in the probability space presents itself as a relational triplet of unit value, a degree of possibility and a degree of probability.

It is useful to keep in mind that the possibility space is composed of possibilistic categories that are associated with all postulated outcomes of a phenomenon where each category is equipped with a possibilistic distribution that captures the degree of possibilistic uncertainty associated with the outcome of a given knowledge phenomenon. Similarly, we expect the categorial conversion of the possibilistic set into probability space to yield probabilistic categories equipped with probability distribution that captures the degrees of probabilistic belief associated with *stochastic uncertainty* of the outcome. The probabilistic epistemic elements appear as complex triplets of the form $\left(\left(x,\mu_{\mathbb{P}}\left(x\right)\right),\mu_{\mathbb{P}}\left(y,\mu_{\mathbb{P}}\left(y\right)\right)\right)$, where $y = \left(x,\mu_{\mathbb{P}}\left(x\right)\right)$ which is referred to as the *fuzzy random variable* with fuzzy (inexact) probability value for each element. The probability space is called the *fuzzy probability space*. The fuzzy random variable may be decomposed into the corresponding units of the triplets. The fuzzy random variable becomes a *classical random variable* with a corresponding *classical probability space* if we assume away the conditions of fuzziness composed of vagueness of language, subjectivity in imagination, ambiguities in communication and others. In the classical system of thought, $\mu_\mathfrak{B}(x)$ in the classical probability space is set to one for each possibilistic epistemic element and hence each $x$ assumes an exact value and meaning with exact probability value under information incompleteness and hence we concentrate on the quantitative volume of information structure to the neglect of its quality. Since $\mu_\mathfrak{B}(x) = 1$, the variable $y = (x, \mu_\mathfrak{B}(x)) = x$.

The probability space is the third pillar in the chain of the knowledge-production process in order to arrive at the epistemic actual. It is, thus, a cognitive construct on the basis of a particular epistemic transformation of information structure to help eliminate some doubts of stochastic belief, and to decide on the degree of probabilistic uncertainty that is tolerable for an acceptance of an event as reasonably true under an explanatory logical system or reasonably actualizable in a prescriptive logical system under some degree of human probabilistic ignorance in the case of the classical thought system, and fuzzy probabilistic ignorance in the case of the fuzzy thought system associated with a particular phenomenon.

The epistemic utility of the construction of the probability space is to help us answer the question of how probable do we expect a possible phenomenon to be logically transformed into the space of the epistemic actual on the basis of quantitatively limited information which will allow cognitive agents to connect from the possible and then to the epistemic actual through the probable [R10] [R10.4][R10.5][R10.11] . The nature of human sense limitations in acquaintance, observation, imagination, recording, remembrance and retrieval imposes on human cognition an information that may be exact or inexact or both providing us with *defective information structure*. Such a defective information structure from the possible must be processed to arrive at a decision on the probable. The probability space and the probabilistic reasoning also help us to identify the probable knowledge element of *what ought to be* that can or will be actualized or can support and explain *what there is* under conditions of a defective information structure. Generally, the probability space is constructed from fuzzy categories that are elements of the possibility space. The elements appear as pairs in the form of categories with possibility distribution. The elements are called possibilistic categories and the variables are called fuzzy variables. The possibility is the result of defective information structure due to vagueness in imagination, intuition, linguistic constructs of meaning, reasoning and communication that characterize our sense data or experiential information structure which may include experimental and non-experimental information sub-structures.

In the construct of the probability space, on the basis of *classical paradigm*, the possibility space is taken as given where any probability set of the phenomenon under investigation is closed by impossibility with probability measure equals to zero, and surety with a probability measure equals to one by definition. No discussions are offered to us as to how the values of zero and one are obtained. In the construct of the probability space, on the basis of the fuzzy paradigm, the possibility space is first constructed where each possibility set for any phenomenon is closed by fuzzy decision-choice action that reconciles the conditions of the surety-impossibility continuum. The fuzzy decision-choice action is used to decide on the tolerance level of epistemic vagueness, ambiguity, and inexactness associated with the elements in the possibility set through fix-level fuzzy algorithms that allow the epistemic possibilistic elements to be transformed into epistemic probabilistic elements.

Generally, two situations must arise in the construct of the probability space as we logically move from the possibility space and traverse between the possible to the probable. One situation involves disregarding the associated degrees of belonging of the possible outcomes in the possibility space, thus converting the fuzzy categories into ordinary categories and hence into ordinary possibility sets where each element is equally possible under identical conditions. The conditions of "equally likely" imply that all elements in the possibility space have equal possibility values that are exact and equal to one, and hence we can neglect the component of the defective information structure due to fuzziness broadly defined to capture the qualitative disposition of information. In this way, we assume away the fuzzy uncertainties, concentrate on defective information structure due to quantitative incompleteness, and transform the complex fuzzy random variable that cannot obey the exact classical laws of thought into a simple classical random variable that can obey exact classical laws of thought. Alternatively, one may carry the fuzzy categories with the associated degrees of belonging to the members in the possibility set into the probability space where outcomes are not equally possible but are weighted by the associated degrees of possibility. The classical way of using Aristotelian reasoning with crisp set is such that the former situation is adopted where the possible elements to be examined for their probable occurrences come to us as un-weighted single variables under a new type of information uncertainty where the elements have no possibility weights. In other words, if $x, z \in \mathcal{P}$, and $x \neq z$, then it is the case that $\mu_{\mathcal{P}}(x) = \mu_{\mathcal{P}}(z) = 1$, which simply means that $\mathcal{P}$ is a crisp possibility set.

In the former situation, the variables used in representing concepts associated with the phenomenon are exact symbolic and linguistic elements that obey the *classical laws of thought* irrespective of the complexities of the categories under epistemic scrutiny for knowledge acceptance of the possible, or for actualization of the possible. The variables are named as *random variables* that are either mathematical or logical objects. They are then manipulated under limited but exact information (stochastic) uncertainty with the classical laws of thought to obtain an exact justification in support of the probabilistic belief that the occurrences will be manifested in the defined environment. The vague aspect of the information structure that produces the fuzzy uncertainty is suppressed in all the sense data or experiential information structure. In this logical frame, one reasons in the classical paradigm where cognition is subjected to the tyranny of the *principle of exactness* that violates every reasonable account of, and adherence to human behavior and linguistic systems with flexible knowing as may be abstracted. This is the problem of *exactness-inexactness duality* in cognition and decision-choice actions. The problem of exactness-inexactness duality is mapped into the representation space of discrete-continuous duality that reflects methodological digital-analog duality in dealing with certainty-uncertainty duality. The problem expands in multiple dimensions as the phenomenon increases in complexity, especially in social, medical, technological and biological systems, requiring new epistemic approaches to representation and the search for justified knowledge items in the knowledge-production process. It is the presence of this problem of the sequence of dualities with the acceptance of the principle of continuum that has given rise to the developments of other knowledge areas such as synergetics, complexity theory and energetics in the contemporary epistemology.

The users of classical paradigm with its mathematics and logic operate like a chess player, with the imagination that they can arrange objects in the epistemological space with precision and ease by the applications of objective rigid rules of thought, just as the chess player operates on the chess board. They fail to account for the fundamental principle of dual motion in that the physical elements on the chess-board retain the same quality and do not exhibit counter qualitative motion and hence, they are easy to represent and manipulate. In practice and over the epistemological space, the epistemic elements alter their forms to incorporate motions of quality and quantity in the chess-board of knowing where the characteristics of the epistemic elements may be elusive due to the simultaneity of qualitative and quantitative motions contrary to the behavior of mathematical objects. This exact rigid determination, in the classical paradigm based on information exactness and operated with exact rules of thought, is replaced by inexact flexible determination with fuzzy covering in the fuzzy paradigm with its mathematic and logic.

If the vagueness that generates fuzzy information is not suppressed in the possibility set as maintained in the fuzzy paradigm, then the latter situation is undertaken where the constructed probability space and the categories of the possibility set will conceptually and computationally appear in triplets as categories of potential elements with corresponding *indexes of possibility* and *indexes of probability* for any given phenomenon. In this way, the potential is continually connected to the possible and then to the probable in order to examine the conditions of the categories of the epistemic actual for collective approval and acceptance by decision-choice action. This situation of the presence of general fuzziness presents a dilemma for the classical laws of thought in human cognition operating with the classical epistemic modules as information processing facility in all areas of knowledge production. Over the epistemological space, the claim of *exact probability* in the classical laws of thought has become contaminated with *vagueness* while the probabilistic reasoning has also become contaminated with *ambiguities*. In this way, the derived results of thought are implicitly inexact in all areas of knowledge production, even though, symbolic exactness in representations, logical exactness in reasoning and mathematical exactness in computations on the basis of exact rigid determination are claimed by the use of the classical paradigm in thoughts. The implicit inexactness is contained in the assumptions on the basis of which the paradigm is manufactured.

With the presence of vagueness and ambiguity, the classical probability space becomes the fuzzy-probability space. The variable representation, in this case, is surrounded by two uncertainties of fuzziness and stocasticity, two interrelated belief systems of possibilistic belief and probabilistic belief, two types of supporting measures of justification of possibility index and probability index and two types of risk of *fuzzy risk* and *stochastic risk* in a composite form called the fuzzy-stochastic risk where we must deal with inexact probability measures. The variable representations in the probability space are not simply random variables but *fuzzy random variables*. The classical probability space with random variable becomes a fuzzy probability space with *fuzzy-random variable*. In this frame, the probability measures are no longer exact, but rather, we encounter either inexact probability [R6.15] [R6.28] [R15.9] [R15.23] or vague probability [R15.8] [R15.12] [R6.24] [R15.26]. The fuzzy-random variables do not obey the classical laws of thought in their manipulations and conceptual combinations on the chess board of knowing. They obey the rules of conceptual combinations of the laws of thought of the fuzzy logic in the fuzzy paradigm. In this way, the fuzzy paradigm allows us to deal with both the quantitative motion that may be external and qualitative motion that is an internal transformation. The fuzzy laws of thought affect all areas of knowledge production in such a way that the claim of principle of exactness in any area of knowledge production is simply by assumption, since the decision-choice processes of every area of thought must deal with the defective information structure and subjective phenomena.

It is here that the logic of mathematics of probability, either in the classical form or in the fuzzy form, becomes helpful in answering the epistemic question of how probable can a possible outcome enter the space of the epistemic actual. It is also here that a problem arises with the classical laws of thought as an attempt is made to account for the qualitative elements of the information structure. In answering these questions, the logicians and the mathematicians using the classical laws of thought discount the vagueness and inexactness associated with elements in the probability space, and ambiguities in reasoning associated with vague logical and mathematical elements in order to arrive at the knowledge acceptance. So far the complete development of fuzzy probability space is ongoing and is the challenge of mathematics that will handle vagueness. It is here that the fuzzy laws of thought with its mathematics bring us some relieve and help to establish a path to follow in the epistemic space whether one is considering explanatory or prescriptive science.

## 3.3   The Probability Space, Explanatory and Prescriptive Sciences

The interpretational elements in the possibility space, as they relate to explanatory and prescriptive sciences, have been discussed. Directly or indirectly, these interpretations also relate to the structures of exact and inexact sciences which are under the subject matter of our epistemic action. They also offer us new and logically fresh alternative views of what constitutes science and non-science and what kinds of assumptions are required for applying different rules of information representations and laws of thought in knowledge development. Similar interpretations of the epistemic elements in the probability space are necessary to understand the chain that connects the building blocks in the theory of the knowledge square from the potential to the epistemic actual passing through the possible and the probable. We must keep in mind that the possible, the probable and the actual are concepts specified in the epistemological space and related to the process of knowing, while the potential is a concept that is related to the ontological space that allows us to initialize the knowledge search dynamics. In general, the theory of the knowledge square involves the questions of how the epistemological space is connected to the ontological space that is taken as the collection of the elements of "what there is" under the knowing process in the epistemic space.

Given the interpretations of the possibilistic categories, what conceptual system do these possibility categories assume in the probability space? The possibility space presents us with possibility sets for each phenomenon where the primary category is established either by empirical or axiomatic foundations or both. Corresponding to the possibility set, regarding a particular phenomenon,  are either 1) a set of explanatory theories with a family of sets of explanatory propositions, or 2) a set of prescriptive theories with a family of sets of prescriptive propositions. To what extent can one believe in these explanatory and prescriptive propositions as probable in relation to the space of the epistemic actual?

The search for answers to the question of how probable do these possible elements that contain knowledge contents leads us to the logical and mathematical construction of a probabilistic belief index or any other belief index to rank the probable elements or find their average behavior and deviations around the average. In the explanatory science, the subject of epistemic analysis is on the examination of competing explanatory theories on the basis of the indexes of probabilistic belief system, derived from the defective information structure due to incomplete information signals associated with the probabilistic categories, and in support of the claimed inference of the derived propositions. The question of how probable involves the distribution of true-false values in duality of the theoretical propositions and their ranking for acceptance of a particular theory. These true-false values are distributed in continuum within the duality. In prescriptive science, the subject of epistemic analysis in the probability space is on the examination of competing prescriptive theories, also on the basis of index of probabilistic belief system derived also from the defective information structure on the basis of volume incompleteness of available information signals given the elements of vagueness. The process is first to transform the possibility set into a probabilistic set and to rank the elements in the probability set for the selection of the most probable, and to establish the average behavior of the phenomenon as well as deviations of other elements around the average.

In both cases of explanatory and prescriptive theoretical systems, the development of the index of probability is related to rationality of decision-choice action regarding true-false claims of the explanatory propositions on one hand and potential-actual transformation claims of the prescriptive propositions on the other. The elements in the probability space, regarding the explanatory science, have no claims to knowledge except claims to the probable acceptances as knowledge. Such acceptances of the explanatory propositions or claims must go through the principles of corroboration, falsification and verification given the possibilistic and probabilistic justified belief systems [R2.9] [R8.27] [R8.56] [R14.45] [R14.49]. The elements in the probability space, in the case of prescriptive science, have no claims to actuality, except claims to probable transformations from the probability space to the space of the epistemic actual, after the implementation of the prescriptive rules.

The claims to probable categorial conversions in the prescriptive science, unlike the case of the explanatory science, are subjected to principles of corroboration, falsification and verification after the applications of the sets of prescriptive rules (for extra discussions see the works in [R2.9]. In the explanatory science, we deal with conditions of justification, corroboration, falsification and verification of knowledge claims. In the prescriptive science, we deal with conditions of justification, corroboration, falsification and verification of outcomes of decision-action claims. This distinction of claims in explanatory and prescriptive theories is very important in the knowledge-production process. Decision-choice actions in both explanatory and prescriptive sciences are constrained by defective information structures that generate fuzzy and stochastic uncertainties, and related to them are the possibilistic and probabilistic beliefs with resulting *fuzzy risk* and *stochastic risk* which are associated with inexact but not exact sciences [R11.21][R11.22]. The fuzzy risk is related to fuzzy conditionality while the stochastic risk is related to stochastic conditionality in the knowledge-production enterprise. They combine to produce an aggregate that is fuzzy-stochastic risk.

The results of the epistemic activities in the epistemological space are that the knowledge structure constructed from the defective information structure irrespective of the area of knowledge cannot take an unquestionable claim to exactness no matter how hard one tries. The point here is that the claim of exactness for some area of knowledge or science is an epistemic fiction whose cognitive and intellectual support is a set of assumptions. It is here that one may link the defective information structure to correctness, accuracy and precision in the knowledge-production system. The universal principle, in the enterprise of knowing, is that all areas of knowledge production are clouded with inexactness. Such inexactness is due to defective information structure that in turn is generated by vagueness in linguistic representation of concepts and ideas, ambiguities in perceptions of the sense data, volume incompleteness of information signals, approximations in human reasoning and faulty communication channels. It is important to point out that the probability index constructed to deal with probabilistic belief also brings conditions of vagueness and ambiguity into the probabilistic reasoning creating a need for concepts of subjective, imprecise, vague or fuzzy probability indexes which have their own logic of analytical construct. The presence of these fuzzy probability indexes in decision-choice actions involving the estimation of exact subjective probability was an intense debate in [R11.22] [R13.26] [R15.8] [R15.9] [R15.26]. This debate may be related to the debate between the camps of the classical and personal probability theories in terms of objective and subjective dispositions.

Furthermore, it may also be pointed out that as the complexity of the system increases, any attempt to impose exactness on the knowledge-production process renders the acquired knowledge irrelevant to decision-choice action. This is a typical case in economics, especially in macroeconomics, where the classical paradigm is used to generate theories about the complex system of human decision-choice actions at both the individual and collective levels. The same situation applies to a number of knowledge areas such as medical sciences, biological sciences, climatology and others. As the complexity of the system of analytical interest increases in dimensionality, the defectiveness of the information structure increases with increasing information vagueness, conceptual ambiguity and information incompleteness, that make it difficult to apply the classical laws of thought and mathematics to derive unquestionably sustainable useful propositions. In all these, we have to deal with the simultaneous presence of quality and quantity with the corresponding qualitative and quantitative motions where the complexity increases with increasing qualitative motion.

In the epistemological space, exactness is the ultimate that we seek in the knowledge-production process given the quantitative volume of information. Its attainment is approached through the process of reducing the fuzziness of the defectiveness in the information structure so as to reduce the possibilistic uncertainty and risk. It requires a reduction in ambiguity and vagueness in representation, a reduction of imprecision of information signals, a reduction of the size of subjective information, an improvement in the logic of reasoning and communications of the known. In all these human activities, there exists an *irreducible core of fuzzy uncertainty* in any area of knowledge production. This irreducible core of fuzzy uncertainty generates an *irreducible core of inexactness*, which in turn produces an *irreducible core of fuzzy risk* for any decision-choice system. Such irreducible core of fuzzy risk is the minimum component of *systemic risk* associated with any decision-choice system whether it is social, physical or non-physical in the knowledge-production process. The desire to derive exact theoretical propositions and to claim exact knowledge items, through the use of classical mathematics and exact reason, forces theories into quantitative straight jackets at the sacrifice of important qualitative characteristics. This is the problem raised in [R14.79] [R14.82] in relation to quality, quantity and time. The point here is that, for the mathematics and exact rigid determination to work, the internal configuration that generates qualitative motion must be assumed to be static and the decision-choice expectations that link the *time trinity* of past, present and future must be static. This point will need some further discussion. It may be pointed out that, in this meta-theoretic framework, possibility relates to quality and probability relates to quantity of information. The concepts of possibility and probability spaces that are discussed here are defined through the concept of the information phenomenon which is then mapped onto the space of the epistemic actual to which we now turn our attention.

# Chapter 4
# The Knowledge Square and the Concepts of Actual and the Space of the Epistemic Actual ( 𝕬 )

We have discussed the foundational blocks of the spaces of the potential, possibility and probability and their relationships to the process of knowing in order to reduce human ignorance in all areas of knowledge. We shall now turn our attention to examine the space of the epistemic actual and discuss the connectedness of this space to the spaces of probability, possibility and the potential (the space of the universal object set) in the reduction process in the knowledge production. The space of the epistemic actual is the final destination of our knowledge journey and the results of the game of knowing. It helps to close the theory of the knowledge square as a continuum process of thought where the chain is never broken, and guarantees that the knowledge production is a never-ending process in the sense that the space of the epistemic actual is also ontologically and epistemologically connected to the potential space through logical dualities and creative tensions. It is here that the principles of corroboration, falsification, verification and acceptance of true or false propositions and knowledge claims acquire meanings. It is this analytical structure of the theory of the Knowledge square that allows scientific and philosophical assessments of meanings embodied in the potential, possible, probable and actual and the corresponding developments of logical paradigms and mathematical structures to deal with them.

## 4.1 The Knowledge Square and the Concept of the Epistemic Actual

To understand the concepts of possibility and actuality and their relationships, they must be referenced to the concepts of the potential and the probability and how they are related to the possibility and actuality in the epistemic process. Similarly, to have a reasonable understanding of the meanings of the concepts of the *possible worlds*, *actual world*, and the world epistemic actual, and their relationships, we must understand the conceptual ideas about the analytical relationships of the potential space, possibility space, probability space and the

K.K. Dompere: The Theory of the Knowledge Square, STUDFUZZ 289, pp. 55–72.
springerlink.com                              © Springer-Verlag Berlin Heidelberg 2013

space of the epistemic actual. We have pointed out that the possibility space for analytical works is smaller than possible-world set and that every element of the possibility space generates a set in the possible worlds by convex combinations where such convex combinations generate combinatorial imaginations in the construct of the possible world set. However, no element in the possible world set can be conceived and generated outside the four spaces. Similarly, no human invention and engineering is possible outside the four spaces of the knowledge square.

From the view point of polarity and duality, with a well structured fuzzy logical and mathematical process, every actuality is a potential in a transformation and every potential is an actuality in the making through a process where the actual fades into the potential, the potential fades into the possibility, the possibility fades into the probability and then back to the actual to complete the knowledge square with further transformational connections to the potential in a continual process without end. The journey through these spaces is the work of epistemic actions with inputs of defective information structures. No segment of the knowledge production can, with comfort, claim exactness of its inquiry due to the presence of the irreducible core of inexactness associated with all human decision-choice actions working through all possible information structures. The best we can claim is varying degrees of exactness from which we can construct a fix-level set by human reason and judgment to establish an ordinary set of exact science from inexact science by fuzzy decomposition. In this way, every element claimed to be exact in science has an irreducible core of inexactness that is accepted as a risk of knowledge failure and ignorance in that science. But this is true in all knowledge sectors. The implication is that any potential knowledge is an open fuzzy set which can be closed only by decision-choice actions on behalf of cognitive agents. The potential, the possible, the probable and the actual are connected by continuous processes that are relationally epistemic, and no two of them appear as parallel as seems to be implied in Buchler [R9.3]. The defining spaces of the potential, the possibility, the probability and the epistemic actual must be seen in terms of quality, quantity and time as key sets of categories of knowing.

For the understanding among the concepts of the potential, the possible, the probable, we shall offer a definition of the concept of the epistemic actual and the space of the actual. Both the concepts of actual and the space of actual are cognitive constructs. The space of the cognitive actual is taken as the space of epistemic reality.

**Definition 4.1.1: The Actual and Reality**

An actual in a conceptual scheme of the knowledge square is that which is cognitively abstracted with epistemic methods from the probability space through a cognitive transformation of the conceptual probability with a justified degree of belief regarding the notion that the epistemic element has been claimed with acceptable methods to be a justified knowledge item, and hence belongs to the epistemic reality which constitutes the model of the cognitive actual.

## Definition 4.1.2: The Space of the Actual

The space of the actual in the conceptual scheme of the theory of the knowledge square is a conceptual subsystem that provides us with a linkage between the probability space and the space of the potential which allows the understanding of the knowledge transformation process in the epistemological space where the *probabilistic epistemic elements* are cognitively believed and hence accepted to be informationally actual as a knowledge item with some degrees of *certainty*. The collection of all elements that are moved from the probability space with justified belief constitutes the space of epistemic actual. The elements of the space of epistemic actual are called the *knowledge elements* and the space may represented as.

$$\mathfrak{A} = \begin{cases} (z)\,|\,z = \left(x, y, \mu_{\mathfrak{B}}\left(y\right), \mu_{\mathfrak{P}}\left(x\right)\right), \left(x, \mu_{\mathfrak{P}}\left(x\right)\right) \in \mathfrak{P}, \\ \text{and } \left(y, \mu_{\mathfrak{B}}\left(y\right)\right) \in \mathfrak{B} \text{ with } \mu_{\mathfrak{A}}\left(z\right) = \left(\mu_{\mathfrak{P}}\left(x\right) \cdot \mu_{\mathfrak{B}}\left(y\right)\right) \in \left[\alpha, 1\right) \end{cases}$$

Where

$$\begin{cases} \mu_{\mathfrak{P}}\left(x\right) = \text{the degree of possibility of knowledge, a measure of possibilistic belief} \\ \mu_{\mathfrak{B}}\left(y\right) = \text{the degree of probability of knowledge, a measure of probabilistic belief} \\ y = \left(x, \mu_{\mathfrak{P}}\left(x\right)\right) \text{ is a possibilistic epistemic variable from the possibility space} \\ \mu_{\mathfrak{A}}\left(z\right) = \alpha \text{ is the maximum tolerance level of risk of knowledge acceptance} \end{cases}$$

## Note 4.1.1

The space of the epistemic actual is equal to or less than the space of cognitive reality which is analytically constrained. It is also a conceptual system that allows us to connect our knowing process of epistemic reality to the ontological space where the epistemic elements are examined and confirmed against the ontological elements of *what there is*, where quantitative and qualitative distinctions of the identities of the epistemic elements in simplicity and complexity are established. The space of cognitive reality is made up of the elements of epistemic actual that are believed to have isomorphic correspondences with some elements of ontological space.

There are two types of possibilistic and probabilistic categories. One type is *cognitive*, in terms of knowing *what there is*. The other type is *natural* and *social* in the sense of categorial conversions through logical substitution-transformation processes. In the former, there is a seed of knowing what the actual and its behavior are. In the latter, there is a seed of the actual in the process of making and becoming. In both cases the actual resides in the probable, the probable resides in the possible and the possible resides in the potential in a continuum. All of these are relationally connected to produce the universal object set that is infinitely closed under the substitution-transformation process in knowing and categorial conversions. In this process, the old actual fades into the potential from which the possible is born to give rise to the seed of the probable that gives birth to the actual as a new candidate to fade into the potential. This is the structure and analytical foundation of the knowledge square and its contribution to the theory of knowledge.

The theory of the knowledge square projects an infinite continuum of substitution-transformation processes in polarity and duality of quantity and quality spaces, with time acting as a neutral element where qualitative and quantitative spaces and time emerge as one in the ontological and epistemological processes whose epistemic connections are the work of information and paradigm of thought. The substitution-transformation process is the destruction of old knowledge and the replacement of the old with a new knowledge of the same phenomenon with different qualitative disposition through forces of creative tension. In this way, the knowledge production and the growth of knowledge are destructive-constructive processes where decision-choice actions operate in the ignorance-knowledge space in a continual process.

From the conceptual system of the knowledge square, a question arises as to what can be said about the *methodology* in the knowledge-acquisition process. The essential proposition from the theory of the knowledge square is that the methodological approach to cognition follows the same path and this path is a subject-matter neutral. The first observation is that the universal object set contains every element in the potential space and is reflected in the ontological system. The elemental contents are taken to be *what there is*. The elements are made up of natural and social categories which contain objects, states and processes that are named in a given language and identified and defined by quality-quantity characteristics at any given time. From the potential space, we may cognitively abstract the social and natural possibility sets, the elemental contents of which, by the very nature of their identifications and categorizations of the linguistic structures, inherit the property of *vagueness* in the possibility space. The vagueness may be associated with *quality* but not necessarily the quantity of information needed for epistemic action. This vagueness is a characteristic of information on all social and natural phenomena in the knowledge-production systems. In fact, it is the property of quality that gives identity to an entity and distinguishes it from other entities in the universal object set as well as creates the conditions for the essential categories to be formed in the epistemological space.

The process of thought formation requires information and information processing which in turn require developing inter-relational and intra-relational structures among the elements in the possibility set in accordance with their defined qualitative characteristics, irrespective of whether the elements belong to a social or natural possibility set. The objective is to move the thought system from the possibility space, where we can examine the sufficiency of the quality of information, to create the probability space where we can examine the sufficiency of the quantity of information, or simply from the possibility set in order to create the probability set. The thought processes in all cases are plagued with *ambiguities* in language and *inexactness* of reasoning requiring subjective decision-choice action for relational clarity. This condition is general to all knowledge sectors whether social or natural. The qualitative characteristics of vagueness, ambiguities and subjective are grouped as *fuzzy phenomenon* with *possibilistic belief system* under dual creative tension. The resolution of this creative tension becomes a decision-choice support to the formation of the possibility set. This possibilistic belief system may be divided into two sub-systems, one of which is associated with natural possibilistic category and the other associated with social possibilistic category. The fuzzy characteristics in the possibility space are carried on to the probability space and disposed of in the space of the actual by decision-choice action. The elements in the Periodic Table are identified not simply by their quantitative characteristics but by their qualitative characteristics. For example, an once of gold and an once of silver are of the same quantity of weight but we know them by the differences in their qualitative characteristics. Inter-categorial differences are established by qualitative characteristics while intra-categorial difference are established by quantitative differences which may or may not be supported by intra-categorial qualitative differences which are defined in terms of aesthetics.

At the level of the probability space, one is confronted not only with the problem of quality of information from the possibility set but with the problem of the quantity of information associated with it given the quality. The probability space will be of no analytical use if we have complete information. The problem of quality of information affects all areas of knowledge production including the epistemic process in nature and society, and hence affects natural, social, medical, behavioral sciences and related ones. The nature of the problem has been explained under the epistemic understanding of the probability space, the conditions of the probability set and the corresponding probabilistic belief system. The epistemic path is such that we may partition the probability set into natural probability set (that is probability set associated with natural phenomena) and social probability set (probability set associated with social phenomena) for analytical convenience, but we cannot get rid of the associated limited information that constrains certainty of exactness. We also cannot get ride off fuzziness that constrains the exactness of certainty. Furthermore, the processing of the probabilistic information to arrive at propositions about the elements in the space of epistemic actual also suffers from the same problems irrespective of whether one is in natural or social sciences. The problem is that the input of the knowledge-production process is of defective-information structures and the information processing is constrained by capacity limitations that lead to approximations by cognitive agents operating in the epistemological space with decision-choice actions.

The thought process in the probability space is also plagued with probabilistic *ambiguities* and reasoning *inexactness* which requires subjective decision-choice actions for relational clarity of the probability values as measures of stochastic beliefs. This is general to all knowledge sectors whether social or natural. From the probabilistic set and with appropriate probabilistic reasoning, we work with knowledge acceptance conditions under *stochastic* and *fuzzy risks* to construct the elements in the *epistemic* reality. This path of sequence is the *methodology* of the knowledge-production process as projected by the theory of the knowledge square. The theory of the knowledge square projects a *universal* principle for cognitive intra-categorial and inter-categorial conversions in every stage in the knowledge-production enterprise. It holds for natural sciences, social science, behavioral science, medical sciences, chemical sciences, economics, psychology political science and other knowledge sectors as well.

The essential features of the theory of the knowledge square are the explanation to the processes through which defective information structures, information representation, paradigms of thought and decision-choice action arise in the epistemological space, as we seek to connect the process of knowing over the epistemological space to the ontological space. The toolbox of the knowledge square in dealing with the essential features is composed of polarity, duality, opposites, continuum, contradiction and the use of fuzzy paradigm in the knowledge-production process. From the conditions of the theory of the knowledge square, we shall offer a definition of epistemology as it is explicated and used in these discussions, and then analyze its relative importance to the development of toolboxes for knowledge search.

**Definition 4.1.3: Epistemology**

Epistemology is a universal thinking framework for knowledge production and the process of knowing (ignorance reduction). It is independent of the content or the area of knowledge sector, its axiomatic and empirical foundations of the primary category, techniques and methods of acquiring knowledge through information processing to create derived categories from the primary categories on the basis of sequences of justified possibilistic and probabilistic belief systems. It is a framework of creating an epistemic actual through the spaces of potential, possibility and probability where the collection of all the primary and derived categories constitute the epistemological space.

The general conditions of epistemology define the space of the knowledge-production game. These conditions are made up of information, irrespective of how it is acquired, and information-processing modules which involve paradigms of thought. We shall now distinguish between epistemology from methodology by a definition of methodology as a subject area of study and as a complement to epistemology.

**Definition 4.1.4: Methodology**

Methodology is a cognitive discipline in the knowledge-production process devoted to a  systematic and logical study of sets of correct and incorrect techniques, methods, rules, procedures and principles of reasoning that may be used within a given paradigm over the epistemological space for knowledge search in various knowledge sectors as may be needed by the nature of their specific subject matters to deal with the possibility space, the probability space and the space of epistemic actual in order to connect the epistemological space to the ontological space through the spaces of the potential, possibility, probability and the epistemic actual. It examines the appropriateness and suitability of all learning and research tools that may vary from basic assumptions, to particular research techniques for different areas of the knowledge-production process given the paradigm of thought. The methodology, therefore, may be viewed as a module for examining an optimal design and application of techniques, methods, rules, procedures and principles of reasoning with a paradigm.

As explicated, epistemology is, thus, a *cognitive frame* that is *synthetic* in its approaches to knowledge production. Conceptually, epistemology defines the space of cognitive activities on the basis of an accepted paradigm for knowledge search, production and knowing. The epistemological space is, thus, general for all knowledge-production systems. Methodology is a too*lbox frame* that is *analytic* in its approaches to knowledge search, knowing and discovery over the epistemological space. Both the cognitive frame and toolbox frame constitute the *epistemic frame* of the knowledge construction-reduction system that is self-organizing, self-contained and self-correcting.   The theory of the knowledge square is presented to deal with the conditions of the knowledge search process and its path to knowing but not to deal directly with specific content of knowledge as a product in cognition. As such, we have consistently used the phrase the knowledge-production process to suggest that we are about the process and its path for knowing and acquiring knowledge.

The theory of the knowledge square is concerned with the broad conception of epistemology, the building blocks of the epistemological space, the nature of the information structure that gives rise to paradigms which give birth to different inter-paradigm and intra-paradigm methodologies for efficient epistemic practice within paradigms. It is a meta-theoretical system on the knowledge production process. It is not concerned with the examinations and developments of specific methods, techniques, rules, principles and others that may be superimposed on the information-knowledge- production process by the specific needs of different knowledge sectors on the knowledge-production process. This path of knowing is general to all knowledge systems, thus projecting a universal principle to knowledge production.

Epistemologically, therefore, *exact science* is not different from *inexact science* except in content, problems and methods of knowledge abstraction and knowing. There are, therefore, no differences in social science, natural science, medical science, behavioral science, biological science, cognitive science, physical science and other areas of the knowledge production over the epistemic space. The acquisition of knowledge items in any of these areas needs information input, a method of information processing, rules of claiming discovery of knowledge, and rules of acceptance of knowledge claims. The methods of acquaintances and information acquisition will vary over knowledge areas. One thing that is common among all these knowledge sectors is that no claim of perfect information (non-fuzzy and complete) can be assured irrespective of the mode of acquaintance except by assumption. All knowledge areas must work with defective information structure (fuzzy and incomplete).   The interactive processes of the knowledge square, cognitive frame, toolbox frame and epistemic frame, as they relate to synthetic and analytical logics in a complex system of the knowledge production process and cognitive agents, are shown in Figure 4.1.1.

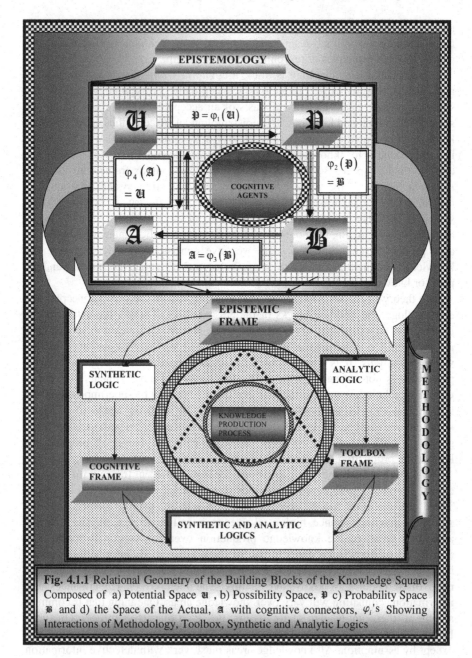

**Fig. 4.1.1** Relational Geometry of the Building Blocks of the Knowledge Square Composed of a) Potential Space $\mathfrak{u}$ , b) Possibility Space, $\mathfrak{P}$ c) Probability Space $\mathfrak{B}$ and d) the Space of the Actual, $\mathfrak{a}$ with cognitive connectors, $\varphi_i$'s Showing Interactions of Methodology, Toolbox, Synthetic and Analytic Logics

Here, given the ontological space, a distinction is made between epistemology and methodology that will allow us to examine the categories of exact and inexact sciences from the position of duality and continuum such that exactness emerges from inexactness through a process. Inexactness is a property of the epistemological space and exactness is the property of the ontological space. The epistemological space is defined by the conditions of the knowledge square given the ontological space. The methodological space is established by pyramidal logic that presents a relational structure of six elemental tools of synthetic and analytical logics in a relational pyramid that is superimposed on another relational pyramid of epistemic, cognitive and toolbox frames in an interactive mode for the knowledge-search activities in the epistemological space. The set of frames constitutes a pyramid. Similarly, the set of logics constitutes a pyramid and their interactions constitute a complex system of knowing.

## 4.2 Epistemic Clarifications of Discretness, Continuum and Reality

Let us turn our attention to provide some clarifications and reflections on the conditions of relationality among some concepts and ideas that have been used in these discussions and will be used in further discussions in the rest of the monograph. The theory of the knowledge square has a number of building blocks that must be related to the development of the general theory of knowledge production. There are two organic building blocks that must be related in the knowledge-discovery process.

### 4.2.1 Clarifications in the Ontological Concepts

First we have the organic building block of the space of *what there is*. The cognitive representation is called the conceptual ontology. The elements of *what there is* are called *ontological elements*. The ontological elements are made up of objects, processes and states that are under continuous transformations. The conceptual place that houses these ontological elements is called the *ontological space*. The collection of these elements that may be accessible to knowing is called the *potential set*. The identities of all the elements in the ontological space are defined by sets of characteristics that are independent and unknown to cognitive agents. The family of the sets of defining characteristics for the identities of the ontological elements is called the ontological information characteristic set.

Each ontological element is defined by a unique set of *ontological characteristics* at any natural transformational period. The unique sets of the characteristics of the ontological elements allow multiple give-and-take relations to create antagonistic, similarity, and differences among the ontological elements. These relationships are established among the elements whether they are aware or not. Besides establishing uniqueness, the relationships and the characteristics allow groups to be formed called *ontological categories*. The essential property of the ontological space is that it is continuous in the sense of analog and in the sense that the ontological characteristics are information flows that are propelled by energy for uninterrupted communications and transformations. This is the *principle of ontological continuum* where relationships and categorial conversions of the ontological elements are continuous and eternal in terms of matter-energy information trinity in the ontological times. The behaviors of the categories obey the natural laws of categorial conversions on the basis of information energy and decision-choice actions under the principle of continuum.

The information that flows from the ontological characteristics is the *ontological information*; the decision-choice actions that lead to continual transformation of the ontological characteristics is the *ontological decision-choice system* and the transformation process is called the *ontological categorial conversion*. The results of interactions of information and energy that provide the forces of categorial conversion is called the *ontological categorial moment* that operates in continuum on the basis of ontological time and space that is expressed in quality-quantity-time space. The quality-quantity-time space with simultaneous interactions of qualitative and quantitative motions may be split into 1) quality-quantity space with transformational dynamics, 2) quality-time space with qualitative motion and 3) quantity-time space with quantitative motion which includes space-time dynamics. The complete ontological space, the universe, is not only continuous but infinitely closed under continual transformations and categorial conversions.

Since the ontological universe is infinitely closed under transformational continuum, it cannot expand. It can only alter qualitative characteristics and rearrange the ontological elements. The idea of conceiving the ontological universe as expanding must also admit the existence of an *empty space* in terms of space-time phenomenon that is not accounted for in our conception of infinity to which an ontological expansion can occur, whether our concept of infinity is explicated as a simple infinity, super infinity or infinitely infinity or infinity of infinities. In this way, our definitional concept of the universe is incomplete if there is an empty space to which it can expand. This is not a discussion on certain claims in the cosmology in relation to the concepts of dark matter, energy and their distribution. Our concern here is the nature of our linguistic representation of what we claim to know and whether such claims are consistent with *what there is*. An infinitely complete universe defined in terms of matter, energy and information cannot quantitatively expand as viewed in terms of space-time phenomena. Individual ontological elements can expand but they do so at the expanse of contraction of other ontological elements. This is also the principle of cost-benefit duality where every beneficial transformation has a supporting cost transformation. As it is seen here, the universe is, however, continually transforming itself in terms of quality that gives meaning to continual creation without end. Some clarifications of the concepts of epistemology and methodology as defined in this monograph are in order.

## 4.2.2 Clarification of Epistemological Concepts

Given the existence of the collection of *what there is* and that the ontological universe is in a continuum, the questions that face the cognitive agents are how the ontological elements can be known and how their behavior can be conceptualized. These questions are existential and different from, but are complemented by the question of how the universe came into being in terms of creation. The first two questions involve the knowing process of ontological existence while the second involves the knowing process of the ontological creation. These questions bring us to the second global building block which is the space of knowing that is constructed by cognitive agents. The space of knowing is the epistemological space that contains the *epistemic elements* which are elements to be known by cognitive agents (what is to be known). From the universal object set of the ontological space and from the epistemological space that defines the environment of the process of knowing, three interrelated spaces are established. They are the spaces of possibility, probability and epistemic actual. The four blocks of the potential space, the possibility space, the probability space and the space of the epistemic actual constitute what defines the knowledge square. The study and the development of their connections into a thought system form the theory of the knowledge square. The logical structure and the conceptual system of thought of the theory of the knowledge square establish a foundation and a universal principle for the development of theory of knowledge. The principle is universal in the sense that it applies and unites all areas of justified knowledge as a complete unified knowledge system [R9.23] [R11.34] [R14.79] [R14.83].

The development of the general knowledge system requires information input from the ontological space through the sending-receiving process of the characteristics of the ontological elements. The information about the ontological elements is ontologically coded as objective information that contains the description of the ontological characteristics in the sense of *what there is* through acquaintances at the primary level of knowing. The sending process from the ontological elements requires energy which is continuous. The cognitive agents receive this ontologically objective information as information signals and subjectively translate them into *epistemological information.* This epistemological information is used to specify the *epistemic elements* and the *knowledge possibility* set which is described by the *epistemological characteristic set* that helps to give identity to the epistemic elements and establish *epistemic categories* in the possibility space. The possibility set waits to be processed into the probability set and then to the set of epistemic actual.

The process of knowing *what there is* is an input-output duality where the ontological information is transformed into epistemological information about epistemic elements which then becomes input into the knowledge-production processes. The knowledge-production process is an epistemic activity undertaken by the cognitive agents to reveal the nature of any epistemic element and its category. It is here that nominalism finds meaning and use in the knowing process. The set of the results of the information-processing activities is the epistemic reality known to be the knowledge as an output that completes the input-output duality which spins the information-knowledge duality. The processing of information is driven by the relational dynamics of *nominalism, constructionism* and *reductionism*. In nominalism, we have information language for the initial decoding of the ontological information into epistemological information which is the subjective information that must be cognitively processed into claimed knowledge. The claimed knowledge characterizes the space of epistemic reality. The information processing is undertaken through the *epistemological decision-choice system* that generates the *epistemological categorial conversion* and *epistemological categorial moment* which provides the epistemic transformational dynamics from the potential space through the possibility space and the probability space to the space of epistemic reality that contains the elements that are claimed to be known. The essential conceptual parallelism is presented in Figure 4.2.1.

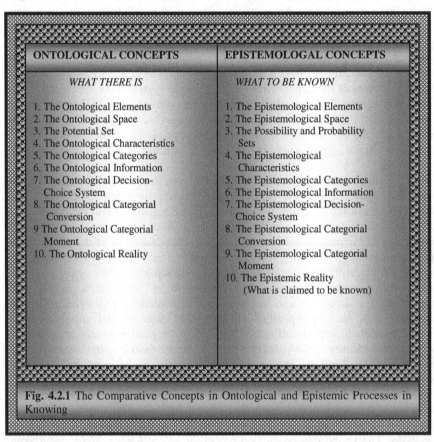

| ONTOLOGICAL CONCEPTS | EPISTEMOLOGAL CONCEPTS |
|---|---|
| *WHAT THERE IS* | *WHAT TO BE KNOWN* |
| 1. The Ontological Elements | 1. The Epistemological Elements |
| 2. The Ontological Space | 2. The Epistemological Space |
| 3. The Potential Set | 3. The Possibility and Probability Sets |
| 4. The Ontological Characteristics | 4. The Epistemological Characteristics |
| 5. The Ontological Categories | 5. The Epistemological Categories |
| 6. The Ontological Information | 6. The Epistemological Information |
| 7. The Ontological Decision-Choice System | 7. The Epistemological Decision-Choice System |
| 8. The Ontological Categorial Conversion | 8. The Epistemological Categorial Conversion |
| 9 The Ontological Categorial Moment | 9. The Epistemological Categorial Moment |
| 10. The Ontological Reality | 10. The Epistemic Reality (What is claimed to be known) |

**Fig. 4.2.1** The Comparative Concepts in Ontological and Epistemic Processes in Knowing

An analytical question regarding information representation emerges in the epistemological decision-choice system. We have argued that the epistemological information is made up of defective information structure composed of quality-quantity characteristics of fuzziness and incompleteness. The representation involves the use of symbolism to create the conditions of nominalism for the information-processing activities and communication that must account for the problem of information defectiveness among cognitive agents. The answer to the question translates into solving the problem of appropriate information representation which is related to formal and informal languages. We have assumed the fundamental ontological continuum of analog representation for the ontological elements, category and space. The justification is that the ontological elements are the work of information and energy which are in continuum. When we come to the epistemological space, what representation must be adopted for the epistemological information? Should nominalism adopt the analog representation or should it adopt the digital representation or both when the representation is in either formal or informal languages that may or may not account for the qualitative aspect of information defectiveness? The nature of the solution to the problem of information representation is extremely important in the working mechanism of the epistemic decision-choice system in the knowledge production, in that it affects the development of paradigms of thought, methodology and the tools of knowledge search within the paradigms.

Our informal languages are developed to account for fuzzy and incomplete characteristics that create defective structure of the epistemological information by combining analog (continuous) and digital (discrete) representations. The problem of appropriate formal representation in terms of logical and mathematical symbolisms has been the first intense debate between the intuitionists on one hand and the logistics and formalists on the other hand. The formalists adhere to the discrete system such as integers to impose exactness which is then related to digital representation and then to discrete processes. The formalist position leads to a search for discrete and digital mathematics and information processing. The intuitionists adhere to the continuum system to incorporate inexactness, such as intervals, which is related to analog representations and then to continuous processes. The intuitionist position leads to a search for methods of interval and analog mathematics and logic of information processing. The formalist school of mathematics and logic finds comfort in the Aristotelian laws of thought with excluded middle and non-acceptance of contradiction as logical value through exact rigid determination. The intuitionist school of mathematics and logic finds comfort in the logic of contradiction in continuum where the principle of excluded middle is rejected. One may refer to the formalist-intuitionist debate on representation and logic [R14.5] [R14.6][R14.14] . Let us keep in mind that the quantitative values in informal language are represented by linguistic numbers such as big, small, tall, short, few, many, little and others which are represented in a continuum where cutoffs are decision-choice determined under inexact flexible determination.

As stated previously, the *fundamental continuum* is a characteristic of the ontological space and the ontological information is also ontologically coded and represented in continuum and carried in continuum without the control of cognitive agents. The debate on information representation in the epistemological space is a methodological one that cognitive agents have control in the decision-choice process. It is about the fundamental principle of appropriate information representation and processing that will provide us with a minimum deviation between ontological and epistemological information for any given phenomenon or a minimum deviation between ontological reality and epistemic reality. Here, we have available to us the use of the methodological continuum or the methodological discreteness or both. The choice of a symbolic representation of information depends on whether one views the universe as composed of digital objects, states and processes or whether one views the universe as composed of analog object, states and processes. In the former, every digital ontological element exists in non-transformational mode with unchanging quality. In the latter, every analog ontological element is in a process of continual transformation with changing quality. One may also choose digital symbolism on the basis of simplicity or analog symbolism on the basis of accuracy at the level of epistemology. The methodological continuum leads to analog and the methodological discreteness leads to digital, both of which belong to cognition and instruments to connect the epistemological space to the ontological space.

The information representation and the study of the ontological continuum may be approached by the methods of either epistemic continuum or epistemic discreteness. Initially, the development of informal languages as information representation tries to combine both to account for both quality and quantity in order to deal with the defective information structure in the epistemological space. For example, the linguistic variables such as red, beautiful, nice and many others that carry qualitative characteristics are defined in continuum or analog. The linguistic quantities such as big, small, tall, and huge and others that carry qualitative characteristics are also defined in continuum. At the level of knowing, the epistemological continuum allows us to account for the part of defective information structure that creates inexactness and hence inexact symbolism and approximation logic. In this case, there is a trade of exactness in simplicity for correctness in complexity. At the level of knowing, the epistemological discreteness allows us to do away with the part of defective information structure that creates inexactness so that exact symbolism and exact logic may be used. In this case, there is a trade of correctness in complexity for exactness in simplicity.

In our attempt to relate the elements in the epistemological space to the elements in the ontological space, should the developments of tools of information representation, processing and understanding be those of inexact symbolism and reasoning, or should they be exact symbolism and reasoning? The two methodological paths for the development of the theory of knowledge presented an intense debate in the field of mathematics and symbolic logic between the intuitionist school operating on non-classical paradigm on one hand and the school of the formalists and the logistics operating on the classical paradigm. The point of disagreement is on the nature of the information that must enter as input into the epistemological decision-choice system for the knowledge production process. Four possibilities in questions are open to us. 1) Should the defective information structure be assumed away where we deal with non-defective information structure composed of exactness and completeness? 2) Should the defective information structure be partially retained where only the incompleteness is allowed while we rid the information input of the component of fuzziness to obtain exact and incomplete information structure? 3) Alternatively, should the defective information structure be partially retained where only the fuzziness is allowed while we rid the information input of the component of incompleteness to obtain inexact and incomplete information structure? 4) Similarly, should the defective information structure be fully retained with fuzziness and incompleteness in order to obtain inexact and incomplete information structure as the input into the epistemic decision-choice system? Questions 1 and 2 are answered in a positive way by the school of the formalists and logics and apply the classical paradigm of knowledge construction, while questions 3 and 4 are positively answered by the intuitionist school and extended to the modern fuzzy-system school, and apply the fuzzy paradigm to information processing toward the knowledge production in all areas of knowing activity.

At the level of epistemic activities over the epistemological space, the following positions are maintained in the construct of the epistemological decision-choice system to transform the epistemic information as input to epistemic reality as the output in these discussions.

1. A space-time continuum is a characteristic of the ontological space. It does not emerge from digital description.
2. Digital description as an exact representation is a characteristic of the epistemological space. It is appropriate in approximation points in the methodological continuum. It follows the analytical representation of exact rigid determination on the principle of the Aristotelian logic.
3. Logical continuum is an enveloping of the discrete process and so also analog is an enveloping of the digital process. The digital (discrete) process is a point-to-point characterization of events in analog (continuum). It is an approximation method.
4. The concepts of minimal length, time, energy, information and how we could observe them are epistemological questions that relate to epistemic reality. They are separate from the ontological reality. In this respect, the fundamental discreteness must be seen in terms of limitations of cognitive agents in their attempts to relate epistemic elements, processes and states from the epistemological space to those of the ontological space in the ontological continuum.
5. The relative conceptual position from the theory of the knowledge square is that the ontological space, the total universe, is infinitely infinite while the epistemological space at any epistemic time, is finitely infinite which is related to different levels of infinity [R14.39][R14.5]. In this respect, the conclusion from the theory of the knowledge square is that the ontological space, the ontological universe, when properly conceptualized, is incapable of expansion in quantitative terms or in space-time relation. Every element is conceived as matter and every matter is energy and information and vice versa where the matter is under constant transformational dynamics that is governed by ontological laws of qualitative motion since it is infinitely self-contained and has no empty space to expand into. The expansions of the ontological elements viewed in space-time processes are quantitatively compensatory while transformations of ontological elements are qualitatively compensatory.

6. The epistemological universe that is either discretely or continuously represented is under continual expansion depending on the activities of cognitive agents in the epistemological space. Alternatively speaking, the epistemological universe has an empty space and hence expansion is possible since the epistemological universe is established by our knowledge system at any epistemic time.

7. The fundamental discreteness may not be associated with ontological activities, the outcomes of which are under cognition but independent of the general human thought system. The principles of continuum, discreteness, analog and digital in the epistemological space must be seen in terms of methodology of knowing through information representation and applications of paradigms of thought. The technique of discreteness allows the use of the classical paradigm to understand the fundamental ontological continuum through the method of sequential problem-solving where x is x and 5 is 5 which is consistent with the classical laws of thought of strict opposites with excluded middle. The digital and discrete methods, in addition to human decision-choice actions, may be used in the construction of the epistemological universe which is substantially smaller than the ontological universe.

8. The epistemological universe is a human construct on the basis of what is claimed to be known in the attempt to know and understand the events in the ontological universe. It is a universe of ideas, knowledge, speculations, dreams, phantom objects, delusions and many others. The ontological universe is made up of matter composed of energy and information where energy flows, information flows and time flows in continuum. The digitization of any of these is taken as methodological simplicity of human inability to handle the simultaneity of qualitative and quantitative continuum in extreme complexity. Methodologically, these flows can be studied with digital process as continuous approximations between intervals. Importantly, a minute, an hour, a day, a month and a year are all human calibrations to understand the flow of time in human activities such as communication, economic production, commerce and others. Similarly, the calibration and unit measurement of volume, length, energy and all others must be viewed in the same cognitive structure and abstraction and must be related to consumption-production duality, selling-buying duality and the phenomenon of socioeconomic exchange. The digital representation and description are consistent with all the flows by decision-choice actions in the epistemological space. Discreteness is an epistemic method of knowing and epistemic reality but not ontological reality. It is an analytical convenience that allows the extensive development of the classical mathematics to be used to derive exact propositions as approximations from defective information structure. Such approximations are usually poor representations of the systems. They may, however, be useful representation for practice with fuzzy-stochastic-risk conditionality. It is here, that the theories of sequential problem solving ill-posed problems and approximations may claim their useful instructions.

9. The question of comparability of ontological space and epistemic method of fundamental discreteness is epistemologically irrelevant since the epistemic process may be viewed as residing in the discrete-continuum or digital-analog duality. In continuum, we find discreteness and vice versa and in analog we find digital and vice versa in the epistemological space. From the view point of the classical paradigm that has taken hold of our knowledge-production process, the only way to represent, describe, model and theorize about the ontological continuum is through the methods of discreteness and digital as a sequential process in understanding and knowing. In this way, the world scientific picture is presented as digital or discrete, and then claimed to be exact without human intervention as it cognitively evolves in stages moving from the potential to the possible and then to the probable and finally to the epistemic actual. This scientific picture as obtained with methods of discreteness and digital is simply a small subset of that which could be obtained with either the methods of continuum or analog or both. The question that arises is simple: When should exactness be accepted by trading off complexity for simplicity and under what conditions should accuracy and relevance be accepted by trading off simplicity for complexity? Here again, we encounter cost-benefit duality and corresponding rationality.

10. From the theory of the knowledge square, an important distinction is made between *ontological reality* and *epistemic reality*. From the established distinction, we can reflect on the question as to whether reality must be purely analog. The ontological reality exists in an ontological continuum and is simply *what there is*. It exists in opposites, polarity, duality and unity in continual transformations through self-adjustments, self-correction and self-integrating. An example is life-death duality that involves all transformation of existence where quantitative and qualitative changes in ontological elements are compensatory in terms of cost-benefit duality. The transformation of life to death is an ontological process in continuum. The epistemic reality is cognitive creation though a methodology and paradigms of thought with their laws of knowing. It may exist in either an *epistemological discreteness* or *digital process*. It may also exist as an *epistemological continuum* or *analog process*.

11. The theory of the knowledge square points to the meta-theoretic position that the epistemic discreteness is a human deficient attempt in simplicity to knowing and understanding of the extreme complexity of the ontological continuum in simplicity. The complexity, synergetics, energetics and informatics exist in digital-analog, discrete-continuous and reality-potential dualities as integrated systems. The applicable paradigm of thought must be the fuzzy paradigm in the epistemological space in order to deal with the defective information structure, linguistic vagueness, ambiguities in reasoning, and information incompleteness where conclusions of exactness and certainty are constructed from inexactness and uncertainty by decision-choice actions of cognitive agents. The methodological discreteness and digital imposes exactness and simplicity by assuming away information

inexactness while the methodological continuum and analog deals with inexactness and complexity to construct conditional exactness and simplicity with analytical claims. This inexactness is a property of all areas of the rooms in the knowledge house.

In the fuzzy paradigm of thought, every symbol and meaning can be represented as a set that shows the degree to which the symbol belongs to the set. Let such a set be $\mathbb{X}$ with a generic element, $x \in \mathbb{X}$. The degree to which $x \in \mathbb{X}$ is represented by a membership characteristic function of the form $\mu_{\mathbb{X}}(x) \in [0,1]$. Thus $\mathbb{X} = \{(x, \mu_{\mathbb{X}}(x)) \mid x \in \mathbb{X}, \mu_{\mathbb{X}}(x) \in [0,1]\}$ is an analog with an interval representation. Digital representation implies that $\mu_{\mathbb{X}}(x) = 1$ or $0$ for exactness with finite members. An analog representation implies that $\mu_{\mathbb{X}}(x) \in [0,1]$ for inexactness which is a fuzzy covering of the digital representation where $\{0,1\} \subset [0,1]$. The digital is simple and more preferred information representation and processing because of exactness and the extensive development of, and general familiarity with the classical mathematics and logic. It, however, loses substantial degree of relevance in representation as the system's complexity increases. The analog is complex and less preferred information representation and processing because of inexactness and the relatively less known and development of the fuzzy mathematics and logic. The inexactness is, however, compensated by increasing degrees of relevance in representation as the system's complexity increases. The work on analog system will reveal new scientific world pictures that are not contained in the digital system which is an approximation of some set elements in the analog system. Digital epistemic reality is a subset of analog epistemic reality which may or may not have correspondence with ontological reality. In fact, there are some mathematicians who have not encountered intuitionist mathematics, how it differs from the currently accepted classical mathematics, the debate on information representation and the rise of fuzzy mathematics. This may also hold in the logical space. This monograph is directed to making explicit the essential differences and the corresponding paradigms of thought for their integrated development and selective use.

# Chapter 5
# Paradigms of Thought in the Fuzzy and Classical Epistemic Systems under Knowledge Production

We have argued from the position of either the theory of complex systems or synergetics in the conceptual system of the knowledge square, that besides the epistemic indifference of the content of the knowledge areas, all areas of knowledge production share in common, the input of defective *information structure*. The defective information structure is composed of fuzzy information deformity that gives rise to *fuzzy uncertainty* and information incompleteness deformity that gives rise to *stochastic uncertainty* from the possibility and probability spaces respectively. In fact, the available toolbox and the logic composed of possibilistic and probabilistic reasoning to deal with these uncertainties are not different in substance and computational structure, except perhaps in style. For the benefit of further differentiations of knowledge areas and new ones that may arise in the future, it is useful to examine the conditions of *inexactness of exact science* and then the *exactness of inexact science*. In this respect, some essential working definitions are needed.

## 5.1 Some Essential Definitions on the Path of Knowing

The examination of exactness and inexactness of sciences requires us to reflect on the *organic paradigms* and their corresponding laws of thought, and how they are related to the knowledge-production process. Let us offer a definition of the concept of the paradigm that is used here. This definition will be a conceptual guide to the understanding of its previous use as well as for further discussions. The ontological space is defined as $\mathfrak{U}$ while the epistemological space, $\mathfrak{C} = (\mathfrak{B} \cup \mathfrak{P} \cup \mathfrak{A})$. Given the ontological space, the epistemological space is composed of the possibility space, the probability space and the space of epistemic actual. All these spaces are defined in terms of the nature of their information structures for knowledge-production, and decision-choice action and the corresponding system of beliefs.

K.K. Dompere: The Theory of the Knowledge Square, STUDFUZZ 289, pp. 73–93.
springerlink.com                                 © Springer-Verlag Berlin Heidelberg 2013

**Definition 5.1.1: Knowledge Item**

An element, $\mathfrak{n}$, in the epistemological space is said to be a justified knowledge item if $\mathfrak{n} \in (\mathfrak{U} \cap \mathfrak{C})$, in that it belongs to the spaces of potential, possibility, probability and epistemic actual with a justified support, $\gamma \in \Upsilon$, where $\Upsilon$ is a family of justification systems. It is said to be an epistemic item if $\mathfrak{n} \in (\mathfrak{U} \cap \mathfrak{B} \cap \mathfrak{P})$, in that it belongs to all other spaces except the space of epistemic actual.

**Note: 5.1.1**

Any justification system is either a rational or non-rational belief system. It is made up of possibilistic belief with a corresponding risk and probabilistic belief with a corresponding risk. In the system of knowledge production, we concentrate on the rational component of the belief system. The collection of all the rational belief systems is called a family of justified belief systems in support of the knowledge claims. This family of the rational belief systems is connected to the paradigms and epistemic conditionality.

**Definition 5.1.2: Paradigm**

A paradigm may be viewed in terms of a set of primary principles and supporting secondary principles to form a closed system in reasoning, in order to pose problems and find solutions, and raise questions and construct answers to them in the search for knowledge through epistemic actions by cognitive agents over the epistemological space of the knowledge-production process. It includes determinations of assumptions about the environment of the knowledge search, information input and representation, logical structure in constructing theories and informed belief structure in claiming the validity of theoretical and empirical conclusions in relation to justifications, falsifications, verifications and corroborations between ontological elements and epistemological elements. A paradigm is an epistemic template which has implied geometry of thinking that is used by a group of cognitive agents for knowledge searches in the knowledge-production process over the epistemological space.

**Note: 5.1.2**

As defined, a paradigm is a cognitive approach in the knowledge-production process given the epistemological space. It defines a particular school of thought (information representation and information-processing module) in the knowledge-production enterprise. It is a decision-behavioral framework of a particular cognitive group over the epistemological space in terms of accepted logical tools for the knowledge production. In particular, it deals with the nature of assumptions about the information structure and the methodological process of transforming information to knowledge by constructing derived categories from a primary category or reducing derived categories to a primary category. In defining methodology, it was pointed out that methodology is directed to the study and analyses of appropriateness of tools and techniques for an optimal application of paradigms.

## 5.2  Reflections on Paradigms and Categories of Laws of Thought in the Knowledge-Production Process

There are two paradigms and the corresponding laws of thought that are of interest to us and essential to the system of understanding that we are constructing in relation to exact and inexact knowledge systems within the theory of the knowledge square. These are the *classical paradigm* with its laws of thought and corresponding mathematics on one hand, and the *fuzzy paradigm* with its laws of thought and corresponding mathematics on the other hand. In the process of discussions, we shall bring into focus the epistemic usefulness of the role of the fuzzy paradigm composed of its laws of thought and mathematics, and what contribution it can make in improving cognition in the knowledge-production process and its enterprise, by placing it alongside the epistemic structure of the classical paradigm in our contemporary rapidly technological change in the *information-decision-interactive processes*. It may be pointed out, that the core argument of the theory of the knowledge square is that our knowledge-production process is an information-decision-interactive process where such information comes to us as defective. The classical paradigm has been in existence since classical antiquity and it is the dominant force in human thinking [R2.12] [R2.18] [R8.11] [R8.17] [R8.56] [R8.62] [R14.20][R14.66][14.97] [R14.113.

The fuzzy paradigm, even though its basic elements can be traced to classical antiquity, is relatively new in its active development. Its understanding role in thinking and use are still under intensive and extensive development; and its use in thought is yet to be fully appreciated. Its logic and mathematical operations present different rules of reasoning that must be learnt by those working with the rules of the classical paradigm. The path of its development may historically be associated with intuitionist logic and mathematics, and their criticisms of formalist mathematics, logic and the use of Aristotelian laws of thought, the principles of excluded middle and the non-acceptance of contradiction [R14.97] [R19.74] in reasoning. The understanding of exactness in thought must be related to the principles of discrete and digital processes, aspects of which were discussed in the previous chapter. The understanding of inexactness in thought must also be related to the principle of continuum and analog processes that we have also discussed.

The two paradigms have similarities and differences that we shall explore in relation to exact and inexact knowledge systems and their uses in the processes of knowing. Additionally, we shall explore the discrete and digital reasoning on one hand and the continuum and analog reasoning on the other, in order to place the discussions on the new demands on contemporary thinking. At the level of similarities, both paradigms have information structure, symbolism, laws of thought, theory of symbolic logic and corresponding mathematics that affect all areas of the knowledge-production process. At the level of differences, there are differences in the information structures, symbolism, the laws of thought, theory of symbolic logic and the corresponding mathematics. The classical laws of thought accept quantity, objectivity, no contradiction in their use, and this position is carried on to the rule of operations of its symbolic logic and mathematics under the *principle of exact rigid determination* which requires exact information structure where quality is assumed away in the epistemological space. It also accepts no inexactness, vagueness, ambiguity and subjectivity. The fuzzy laws of thought accept contradiction as an inherent part of human thinking, and this position is carried on to its symbolic logic and mathematics which can handle both exact and inexact information structures. It also accepts inexactness, vagueness, ambiguity, quality and subjectivity as essential components of the knowledge production in the epistemological space.

These paradigms create the conditions of the cognitive paths of reasoning in connecting *what there is* to *what is known* from the potential to the possible to the probable and to the epistemic actual as seen from the epistemic geometry of the knowledge square. The two paradigms are viewed in terms of toolboxes of reasoning, given the capacity of human cognition, in order to connect some elements in the epistemological space to some elements in the ontological space. The space for knowledge production is complex with categorial differences and similarities. The process of knowing works through information, language and epistemic rationality. The language is constructed on the basis of acquaintances with the ontological elements and epistemic rationality that allows us to accept *what there is* as a satisfying identity condition of knowledge. The nature of the path of knowing depends on how we impose, accept and deal with varying degrees of complexity in the epistemological space given the path of knowledge production.

There are two paths which are available to use in dealing with the complexity in the defective information structure in the process of knowing. One path of knowing imposes discreteness, digitization, exactness and objectivity in the information representation in a given language structure. This path presents simplicity in complexity, and then develops logic with excluded middle that accepts no contradiction as valid, in order to deal with the simplified exact information space where contradictions are abstracted out of the decision-choice process. The other path of knowing accepts the transfer of the ontological continuum, inexactness and subjectivity in information representation in any given language structure. This path retains the complexity and the continuum, and constructs a logic that accepts contradiction in its truth value to deal with the complex information space.

The exact information representation corresponds to the classical laws of thought that exclude the continuum phenomenon and impose conditions of excluded middle in an analytical and synthetic reasoning by merely dealing with the endpoints of the continuum in the knowledge search. The inexact information representation corresponds to the fuzzy logic (infinite logic) that affirms continuum and develops thought in an analytical and synthetic reasoning that incorporates the classical endpoints and the fuzzy continuum in complex systems on the basis of decision-choice dynamics. The relative positions of the two paradigms and the corresponding assumptions about the information structure are presented as relational geometry in Figure 5.2.1. It may be pointed out that discretization and digitization are requirements for imposing exactness, which, is in turn a requirement for the use of the classical paradigm leading to exact rigid determination in all computable systems. Continuum and analog are requirements for the presence of inexactness, which, in turn is a requirement for the use of the fuzzy paradigm leading to inexact flexible determination on the basis of decision-choice system.

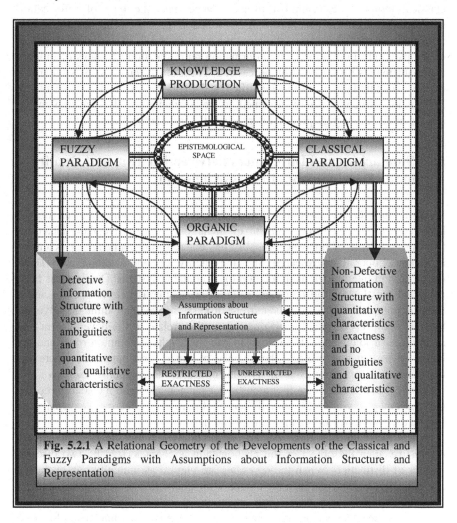

**Fig. 5.2.1** A Relational Geometry of the Developments of the Classical and Fuzzy Paradigms with Assumptions about Information Structure and Representation

The understanding of these two paradigmatic approaches to the knowledge-production process within any given language requires a critical reflection on the structure of linguistic categories and on the forces of substitution–transformation processes, intra-categorial and inter-categorial conversions. All these relations and approaches to knowing and understanding of the relational structure between ontology and epistemology must be seen in terms of analytical tools of dualism, duality, excluded middle, continuum and unity of the knowing process. The analytical tools of opposites, dualism, duality, excluded middle, continuum and unity must be related to continuous, discrete, analog and digital in order to understand and appreciate the relative competing nature of the two paradigms in terms of the reasoning toolboxes that they offer. The fuzzy toolbox and the classical toolbox are fashioned in accordance with the type of information structure assumed in the knowledge-production process. The nature of the information structure assumed dictates the structure and form of the development of the paradigm in relation to its information representation and laws of thought and mathematics for information processing. Thus, the differences and similarities of the two paradigms may be identified and explained with the nature of the information structure that is assumed to be available for the decision-choice actions in the knowledge-production process. Let us examine them.

## 5.3  Similarities and Differences between Dualism and Duality in the Development of Laws of Thought

Dualism and duality have been discussed intensely in all branches of the knowledge-production process. Both of them relate and derive their meanings from the principle of opposites. Duality has entered into mathematics, sciences, economics, operations research and the theory of optimization with some differences in interpretations in relation to computable systems. Dualism has been restricted mostly to philosophy and religion and some aspects of social science such as political science. In our current discussion, we shall limit ourselves to logical conditions of duality and dualism as they relate to the development and acceptance of knowledge. Both dualism and duality have common roots and are based on a dual that projects the essence of binary and opposites, where the opposites may be viewed, either as both opposites are mutually exclusive and collectively exhaustive regarding a particular phenomenon and a category, or, the opposites may be viewed as mutually non-exclusive and collectively exhaustive in logical derivatives regarding the same phenomenon.

When the opposites are viewed as mutually exclusive and collectively exhaustive, then the object or the phenomenon is characterized by two discrete entities of distinction. When the opposites are viewed as mutually non-exclusive and collective exhaustive, then the object or the phenomenon is characterized by a continuum of opposites which may be mutually indistinguishable at the same points. The idea here is that every phenomenon or entity in the universal object set is characterized by opposites. These opposites may be viewed in terms of negative (cost) or positive (benefit) characteristics that provide identity to each element. We shall use these opposites to place a distinction and similarity between dualism and duality. From the viewpoint of knowing, every ontological element contains negative and positive characteristic sets that retain its identity in both ontological and epistemic spaces. The epistemological characteristic set may deviate from the ontological characteristic set as viewed in terms of the information content.

### Definition 5.3.1: Characteristic Set

The set, $\mathbb{X}$ is said to be a universal characteristic set if there are subsets, $\mathbb{X}_C \subset \mathbb{X}$ that may be assigned to any of the elements in the universal object set, $\Omega$, for identity and distinction of category $\mathbb{C}$ to which identical elements belong. The characteristic set, $\mathbb{X}_C$ is said to be category-specific and defines the essential attributes of the classification of the category, $\mathbb{C}$.

### Note 5.3.1

For the purpose of verifying knowledge, the category and characteristic set may be identified as ontological, $\mathfrak{U}$ and epistemological, $\mathfrak{E}$, whose elements must be compared for knowledge acceptance or rejection. Ontological category is $\mathbb{C}_U$, Ontological characteristic set is $\mathbb{X}_U$, Epistemological category is $\mathbb{C}_E$ and the Epistemological characteristic set is $\mathbb{X}_E$. In the ontological space, the characteristic set of each category may then be identified as $\mathbb{X}_{C_U}$ while in the epistemological space, the characteristic set of each category may be identified as $\mathbb{X}_{C_E}$. Let us keep in mind that each ontological element and each epistemic element has the same characteristic set of the category it belongs. In this way, a completely true knowledge requires an equality between ontological identity and epistemic identity of elements such that, $\mathbb{C}_E = \mathbb{C}_U$, and hence the epistemic category is the same as the ontological category. A partial knowledge is then defined to exist if, $\mathbb{C}_E \subset \mathbb{C}_U$. Linguistic representations and symbolism of the information of the characteristics project differential nature of these characteristic sets over the epistemic space. The development of the theory of statistics is to develop the framework, techniques and methods to examine closeness of $\mathbb{C}_E$ to $\mathbb{C}_U$ through estimation and hypotheses testing. From the universal object set to the probability space, the main concern in the knowledge production process is to examine the *information content* of epistemic variables. From the probability space to the space of the epistemic actual the main concern shifts to the examination of the knowledge content of the epistemic variables. In the former, the test for acceptance is on the credibility of the information content in terms of quality and quantity, while in the latter the test of acceptance is on the knowledge content for acceptance in terms of ontological-epistemological knowledge distance (gap).

**Proposition 5.3.1: Existence of Categories**

Every element in a category exists and is defined by its opposites. Its identity is specified by a non-empty negative characteristic set, $\mathbb{X}_C^N \neq \varnothing$, and a non-empty positive characteristic set, $\mathbb{X}_C^P \neq \varnothing$, such that $\mathbb{X}_C = \left( \mathbb{X}_C^N \cup \mathbb{X}_C^P \right)$ and $\#\mathbb{X}_C^N \lesseqgtr \#\mathbb{X}_C^P$ in order to establish the identity of the element and its category in the ontological space.

**Note: 5.3.2**

As it has been previously pointed out, the negative and positive characteristic sets that divide a unit into two opposites and unite them into unity may be taken as discrete entities of attributes with nothing in common that define the qualitative essence of the elements and the category to which they belong. They may also be taken as residing in a continuum that expresses a smooth transition between the extremes to establish the identity of the elements through linkages. The linkages of the negative and positive characteristic sets are through relations that must be established. These relations are complementary, supplementary, give-and-take (reciprocity) and others. The relations may change as time proceeds and transformations of characteristics take place. The essential elements of thinking are that both negative and positive characteristics interact in fulfilling and supplying something that both the negative and the positive lack, to ensure their mutual existence and the survival of the element in the category of their residence.

It is this relationality that ensures the quality identity and integrity of the elements for distinction from others in the epistemological space through informational acquaintance. The relationality of the positive and negative subsets of characteristics is essential in defining the concepts of dualism and duality as they relate to the paradigms with the developments of their logics and corresponding mathematics. Let us keep in focus that every element in the universal object set presents to the cognitive agents both cost and benefit or negative and positive characteristic sets as seen in terms of usefulness and harm. Here, we may specify ontological and epistemological negative characteristic sets respectively as $\mathbb{X}_{C_U}^N$ and $\mathbb{X}_{C_E}^N$, and the ontological and epistemological positive characteristic sets as $\mathbb{X}_{C_U}^P$ and $\mathbb{X}_{C_E}^P$. Since the ontological space is given, we shall concentrate on the epistemological space.

**Definition 5.3.2: Dualism**

Dualism is a conceptual system of representing and thinking about a categorial element with an internal self which is divided into two opposing parts without internal relations except that of relative definition.  In dualism, we have $\mathbb{X}_C = \left( \mathbb{X}_C^N \cup \mathbb{X}_C^P \right)$ with $\left( \mathbb{X}_C^N \cap \mathbb{X}_C^P \right) = \varnothing$ in specifying the condition of excluded middle with $\mathbb{C}$ as its category of belonging.

**Definition 5.3.3: Duality**

Duality is a conceptual system of representing and thinking about an element in the universal object set with an internal self which is divided into two opposing parts with continual interactions in a continuum for their mutual existence. In duality, we have $\mathbb{X}_C = \left( \mathbb{X}_C^N \cup \mathbb{X}_C^P \right)$ with $\left( \mathbb{X}_C^N \cap \mathbb{X}_C^P \right) \neq \varnothing$ that specifies non-excluded middle for continuum with $\mathbb{C}$ as its category of belonging.

**Definition 5.3.4 Continuum**

A Continuum is a conceptual system of representing and thinking about an element in the universal object set with an internal self which is divided into two opposing non-separable inter-supporting negative-positive duality, such that every negative characteristic has inseparable supporting positive characteristic and vice versa that defines its identity and maintains its integrity in a unified whole.

**Note: 5.3.3**

As defined, the duality contains dualism as extreme cases in reasoning. Dualism is, thus, a subset of duality in the true-false space. This leads to the statement that for every proposition of dualism there is a propositional covering in duality. However, not all propositions in duality have correspondence in dualism. Alternatively stated, by the principle of continuum, every true-false proposition in the classical laws of thought has a proposition that covers it in the fuzzy laws of thought. Viewed in terms of mathematical thinking, if we construct any set $\mathbb{A}$ of a phenomenon on the principle of classical laws of thought, and then construct a set $\mathbb{B}$ of the same phenomenon on the principle of fuzzy laws of thought, then $\mathbb{A}$ is always contained in $\mathbb{B}$ $\left( \mathbb{A} \subset \mathbb{B} \right)$. We shall speak of logical dualism and duality.

As stated, in the fuzzy laws of thought, every true proposition has an inseparable false proposition as its support and every false proposition has a true support proposition. Translated in the logic of sets, every negative (false) characteristic set has a positive characteristic set that mutually define their individual identities in continuum and unity.

## 5.4 Exactness, Inexactness and the Classical Laws of Thought

From the definitions of dualism and duality, the classical laws of thought apply to dualism but not to duality as we have defined them. In logical dualism, we work in exact spaces of potential, possibility and probability that have no room for inexactness of any kind in the knowledge-production process towards the space of epistemic actual. When one accepts the descriptive notion of dualism where $\mathbb{X}_C^N$ and $\mathbb{X}_C^P$ exist as independent characteristic sets, then the assignment function from $\mathbb{X}_C$ into $\mathbb{X}_C^N$ and $\mathbb{X}_C^P$ may be specified as a membership characteristic function $\mu_{\mathbb{X}_C^N}(x)$ and $\mu_{\mathbb{X}_C^P}(x)$ with a structure:

$$\mu_{\mathbb{X}_C^N}(x) = \begin{cases} 1 \text{ if } x \in \mathbb{X}_C^N \text{ and } x \notin \mathbb{X}_C^P \\ 0 \text{ if } x \notin \mathbb{X}_C^P \text{ and } x \in \mathbb{X}_C^N \end{cases}, \text{ Negative Characteristic Set} \quad (5.4.1)$$

Similarly, we can specify:

$$\mu_{\mathbb{X}_C^P}(x) = \begin{cases} 1 \text{ if } x \in \mathbb{X}_C^P \text{ and } x \notin \mathbb{X}_C^N \\ 0 \text{ if } x \notin \mathbb{X}_C^N \text{ and } x \in \mathbb{X}_C^P \end{cases}, \text{ Positive Characteristic Set} \quad (5.4.2)$$

In the classical situation, the partition of the characteristic set of every epistemic element is said to be crisp and exact and the dualism coincides with the non-interactive opposites. The fundamental structure of thinking within the classical paradigm is composed of the toolbox of dualism, non-interactive opposites and the classical laws of thought with the corresponding mathematics. These together define the classical exact symbolic reasoning, exact rigid termination and exact computable systems which apply to all areas of the knowledge-production that use the classical paradigm. The problem with this approach to knowledge production is that quality and subjectivity are excluded as essential elements on the path of knowing. In the classical paradigm, certainty and uncertainty are introduced as belonging to the exact space of thought, where uncertainty is related to limited information and measured in the exact probability space with an assumed exact possibility set, all of which tend to produce the *classical epistemic rationality* in the knowledge-production process. The epistemic toolbox defined in terms of pyramidal geometric structure is then imposed on the information-knowledge structure as presented in Figure 5.4.1.

It is important to note that in the classical paradigm of thought, we deal with quantity and time where information is assumed to be exact and full for the *certainty space* and exact and limited in the *uncertainty space*. Ambiguities, vagueness and subjectivity play no essential role in the analytical structure. The failure to discuss the problems of vagueness, ambiguity and subjectivity as part of the internal structure of thought points to an important limitation of the classical paradigm as a general information processing module [R14.6] [R14.4] [R19.3] [R19.4] [R19.58]. The problem of inadequacy of the classical system of thought increases in difficulty in the information-knowledge representation and thought if vagueness, ambiguities and subjectivity are not only related to qualitative phenomena and time, but they are dominant characteristics of the epistemic variable. This situation moves us from the *quantity-time* space to *quality-quantity-time space* with an increasing complexity that creates increasing inexactness.

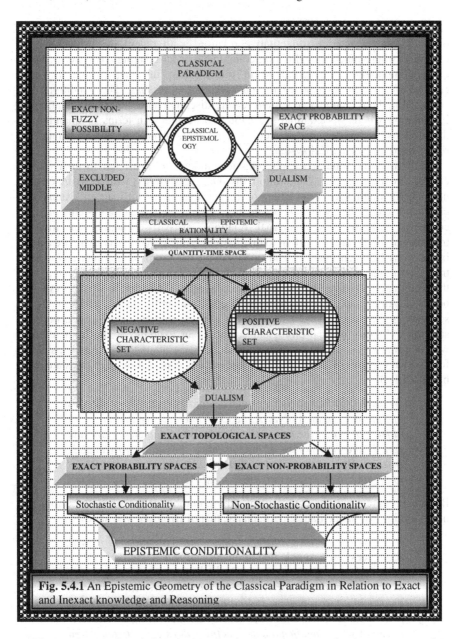

**Fig. 5.4.1** An Epistemic Geometry of the Classical Paradigm in Relation to Exact and Inexact knowledge and Reasoning

The presence of such inexactness in the knowledge-production process strips off the logical potency of the classical paradigm especially as a universal system of thought in dealing with the science of transformations, complexity, synergetics, energetics and social change. The theory of approximations, as developed within the classical paradigm, is an artificial grafting that is still applicable in the quantity-time space. Within the classical quantity-time space are developed the classical mathematics and algorithmic thinking where the elements within opposite are restricted to quantitative relations with regard to correlation and causation where much thought is on computable exact quantitative systems. Such exact computation is referred to in our modern information age as *hard computable system* as opposed to s*oft computable systems*. The conditions for constructing hard computable system require discrete and digital representation of information where we examine precision and accuracy in terms of quantitative dispositions. Linguistic values such as large, tall, plenty and others cannot be represented in the classical system.

In the quantity-time space, the classical thought is restricted to quantitative changes by reducing the system into individual parts (digits) with an assumed and unchanging qualitative disposition of the system. It is difficult to extend the classical logic to deal with self-organizing systems, self-exciting systems or self-correcting systems where internal relations of parts generate substitution-transformation dynamics leading to qualitative changes, and where there are active interactions between quantity and quality with the progress of time; and where transformations are induced from within and by internal controllers in such a way that the quantitative representation is an inadequate description of the system's behavior. An example of complexity with quality-quantity-time structure in social sciences is the social system itself composed of political, legal and economic structures. In the medical-biological sciences, an example is the human body where hormones act as information-knowledge transmitters to the brain subsystem for internal controls. The inexactness of science is captured through the statement by Max Black:

> *While the mathematician* [formalist] *constructs a theory in terms of "perfect"* [exact] *objects, the experimental scientist observes objects of which the properties demanded by the theory are and can, in the very nature of measurement, be only approximately true. As Duhem remarks, mathematical deduction is not useful to the physicist if interpreted rigorously* [R19.22p. 427].

Black's statement may be complemented by a number of positions and criticisms in the intuitionist mathematics, particularly, that of Brouwer who states:

> *It is well to notice in this connection* [isolation of causal sequences and supplementation of human activity] *that a natural law in the statement of which measurable magnitudes occur can only be understood to hold in nature with a certain degree of approximation; indeed natural laws as a rule are not proof against sufficient refinement of the measuring tools....The question where mathematical exactness does exist, is answered differently by the two sides; the intuitionist says: in the human intellect, the formalist says: on paper....*
> *It is true that from certain relations among mathematical entities, which we assume as axioms, we deduce other relations according to fixed laws, in the conviction that in this way we derive truths from truths by logical reasoning but this non-*

> *mathematical conviction of truth or legitimacy has no exactness*
> *whatever and is nothing but vague sensation of delight arising*
> *from the knowledge of the efficacy of the projection into nature*
> *of these relations and laws of reasoning* [classical]. *For the*
> *formalist therefore mathematical exactness consists merely in*
> *the method of developing the series of relations, and is*
> *independent of the significance one might want to give to the*
> *relations or the entities which they relate. And for the consistent*
> *formalist these meaningless series of relations to which*
> *mathematics are deduced have mathematical existence only*
> *when they have been represented in spoken or written language*
> *together with the mathematical-logical laws upon which their*
> *development depends , thus forming what is called symbolic*
> *logic* [R14.14, pp. 77-79].

In the same analytical discussion, Brouwer points out the need for the formalist mathematicians to avoid the use of linguistic variables (fuzzy variables in the current usage) in their analytical developments. Linguistic variables carry with them the qualitative characteristics of the essence of epistemic entities leading to vagueness in representation, ambiguities in reasoning, subjectivity in the interpretations of the conclusion and decision-choice action with respect to the degree of exactness that is associated with the thought system. He, further, raises some important questions of acceptability of systems of symbolic representation of information and the relative preference for non-contradiction over contradiction in truth values. He states:

> *Because the usual spoken or written languages do not in the*
> *least satisfy the requirement of consistency demanded of this*
> *symbolic logic, formalists try to avoid the use of ordinary langue*
> *in mathematics.... Why certain systems of symbolic logic rather*
> *than others may be effectively projected upon nature....Not to*
> *the mathematician, but to the psychologist, belongs the task of*
> *explaining why we believe in certain system of symbolic logic*
> *and not in others, in particular why we are averse to the so-*
> *called contradictory systems in which the negative as well as the*
> *positive of certain propositions are valid* [R14.14, p79].

Fundamentally, the analytical concerns of Brouwer and Black in the knowledge-production process must be seen in terms of the differences in the fundamental assumptions in the epistemological space on the basis of which representation of information may be constructed and the nature of reasoning may also be imposed on the representation for truth acceptance and non-acceptance. These concerns must not be carried to the ontological space whose elements are taken to satisfy the conditions of the law of identity where *what there is*, is *what there is*. All logical schools of thought and the corresponding paradigm seem to accept this law of identity that unites them. They, however, are distinguished and separated by their manner of treating contradictions in true-false acceptance of propositions. The classical paradigm accepts no contradictions in the true-false acceptance of statement, propositions and conclusions. In this way, a statement can be proven to be false by showing a simple contradiction in the sense that it contains the elements of the opposites. Every true statement is free of its opposite in the classical paradigm. This approach is used in proving validities of theorems in logic and mathematics.

The non-classical paradigms, such as the fuzzy paradigm, accept the presence of contradiction as part of reasoning to decide on truth and falsity in propositions, statements and analytical conclusions. In this respect, a statement or a proposition cannot simply be proven false or true by the presence of a contradiction. Any contradiction in a statement, proposition and conclusion is resolved by balancing the relational struggle of true-false duality in continuum which proceeds in doing away with the classical law of excluded middle and replacing it with the law of continuum. One may also refer to the works in [R14.6][R14.14][R14.31]. With these questions and critiques about the fundamental assumptions of the classical paradigm, let us turn our attention to the development of the fuzzy paradigm with the corresponding logic and mathematics. It may be pointed out that the analytical work with the classical paradigm may lead some researchers to try to prove that the universe is discrete and digital rather than continuum.

## 5.5  Inexactness, Exactness and the Fuzzy Laws of Thought

Some current areas of knowledge and emerging ones cannot be fully expanded on the basis of the classical paradigm composed of its laws of thought and mathematics. The paradigm establishes the boundary of problems and questions that can be dealt with.  Among such areas are synergetic science, complexity science, the science of social management, macroeconomics, the science of self-organizing systems, environmental sciences, systemicity and others. The conceptual systems of these areas have one thing in common; they are complex with quantity-quality interactions in time and over time. They are defined and specified in quality-quantity-time space with dynamic qualitative relations among parts where the identity of each individual element, in the universal object set, alters its quality with separate parts in disunity, and when the separate parts are brought into unity, the identity of the element may assume a different quality as it is conceptualized in the epistemological space. For example, if a living being is cut into parts, it is transformed from life to death and when the parts are brought together, it is transformed into a dead body. No human part, for example, can accomplish a task in isolation from the body system. All these systems reflect conditions of energetics and synergetic complexities in such a way that epistemic objects exist in an inexact space whose analytic and synthetic understanding requires a logico-mathematical space, where we cannot do away with vagueness and ambiguities that require subjectivity in judgment as we move over qualitative-quantitative states with time in motion.

In this epistemic frame, the fundamental classical assumptions are replaced by fundamental fuzzy assumptions about the epistemological space such that the connected, the separate, the continuous, the analog the discrete and the digital are united to give rise to the logical continuum with a cutoff process, defined in terms of exactness and truth, is decision-choice determined in the intellect, where every epistemic category and set of characteristics of epistemic element are human constructs. This is where the fuzzy paradigm enters and where fuzzy mathematical objects exist from the cognitive world of epistemic objects that are independent of the thinking of cognitive agents, who simply obey the conditions of exact information representation and the classical laws of thought.

### 5.5.1   How Do the Fuzzy Laws of Thought Apply?

As stated, the fuzzy laws of thought apply under the system of dualities in addition to dualism, where dualism is conceptually contained in duality in the set-theoretic thinking under the principle of logical continuum. The true-false claim proceeds from the conceptual structure of duality. The fuzzy laws of thought may be viewed in terms of the postulate that every element in the universal object set or in the ontological space is defined by their characteristic set, $\mathbb{X}$. When one accepts this defining structure of duality where $\mathbb{X}_C^N$ and $\mathbb{X}_C^P$ exist as interdependent negative and positive characteristic sets, then the assignment function from $\mathbb{X}_C$ as the characteristic set of the category, $\mathbb{C}$, into either $\mathbb{X}_C^N$ or $\mathbb{X}_C^P$ may be specified in terms of membership functions $\mu_{\mathbb{X}_C^N}(\bullet)$ and $\mu_{\mathbb{X}_C^P}(\bullet)$ in the thought process. The complexity of such specification requires us to bring into the analytical process, the essential differences between duality and polarity. Duality and poles are interdependent in their existence. In each pole of a polarity there resides a duality that provides the pole's identity. In this respect, the assignment characteristic function for the negative and positive poles through the membership functions may be written as:

POSITIVE POLE OF $\mathbb{X}_C$

$$\left[\mu_{\mathbb{X}_C^P}(x)\right]^P \begin{cases} \in [0,1]^P, \text{ if the pole is positive} \\ \text{with} \\ \mu_{\mathbb{X}_C^N}(x) = \left[1 - \mu_{\mathbb{X}_C^P}(x)\right] \in [0,1]^P \end{cases} \qquad (5.5.1.1)$$

NEGATIVE POLE OF $\mathbb{X}_C$

$$\left[\mu_{\mathbb{X}_C^N}(x)\right]^P \begin{cases} \in [0,1]^N, \text{ if the pole is negative} \\ \text{with} \\ \mu_{\mathbb{X}_C^P}(x) = \left[1 - \mu_{\mathbb{X}_C^N}(x)\right] \in [0,1]^N \end{cases} \qquad (5.5.1.2)$$

The logic of the duality is such that in the positive pole relative to the negative pole we have:

$$\left[\mu_{\mathbb{X}_C^N}(x)\right]^P = \left\{\left[1-\mu_{\mathbb{X}_C^P}(x)\right]^P \in [0,1]^P\right\} \neq \left[\mu_{\mathbb{X}_C^N}(x)\right]^N$$

$$\left.\begin{array}{l}\\ \text{That is, } \left[\mu_{\mathbb{X}_C^N}(x)\right]^P \neq \left[\mu_{\mathbb{X}_C^N}(x)\right]^N \text{ and } \left[\mu_{\mathbb{X}_C^P}(x)\right]^P \neq \left[\mu_{\mathbb{X}_C^P}(x)\right]^P\end{array}\right\} \quad (5.5.1.3)$$

Every polarity $\mathbb{P}$ may then be specified in terms of membership functions as:

$$\mathbb{P} = \left\{(x,\mu_{\mathbb{P}}(x)) \mid x \in \mathbb{X}, \ \mu_{\mathbb{P}}(x) = \left[\mu_{\mathbb{X}_C^P}(x)\right]^P \vee \left[\mu_{\mathbb{X}_C^N}(x)\right]^P, \mathbb{X} = \left(\mathbb{X}_C^N \cup \mathbb{X}_C^P\right)\right\}$$

$$(5.5.1.4)$$

Linguistically, the membership negative (positive) characteristic set in the positive pole is not the same as the membership negative (positive) characteristic set in the negative pole. The defining epistemic condition is such that the polarity and duality are not defined by crisp sets nor do the sets coincide with each other, but they exist in interactive and interdependent relations that define their mutual existence as well as the identities of the elements in the universal object set.

The interactive structures of polarity and duality generate the system's complexity and synergetic relations with quality-quantity interactions, where the knowledge-production process imposes upon the cognitive agent the need to make judgments, and externalization of the cognitive agent is impossible, because the knowledge-production process is embedded in an inexact epistemological space due to the presence of the defective information structure. In this case, the partition of the information space is said to be vague, fuzzy and inexact. The epistemic elements unlike the ontological elements are inexact. The fundamental structure of information processing through thinking is composed of the *toolbox of duality, opposites, polarity, continuum and fuzzy laws of thought with corresponding logic and mathematics* that define the fuzzy paradigm. The advantage of the fuzzy paradigm is its capacity to allow quantity, quality, subjectivity, learning and the correction process to be integrated into cognition in a manner that admits of errors of judgment in reasoning with further allowances for the information-knowledge production system to retain its position as self-exciting, self-organizing and self-correcting towards exactness and perfection. This is essential for the study of social dynamics, cybernetics of physical and biological systems, climate science and quantum phenomena that the classical exact reasoning and rigid determination fail to capture.

In the fuzzy paradigm, certainty and uncertainty are introduced as belonging to the inexact space. Certainty is defined in the fuzzy space in the sense of relative contents of vagueness and ambiguity, and where uncertainty is related to limited fuzzy information and measured in fuzzy probability space with defined possibility sets. All of this tends to produce *fuzzy rationality* in thought that incorporates boundedness of rationality through the classical extremes and fuzzy continuum phenomenon. The epistemic process is to do away with the *principle of excluded middle* and to replace it with the *principle of continuum*. The epistemic toolbox defined in terms of pyramidal logical structures is then imposed on the information-knowledge space that passes through the quality-quantity-time space as represented in Figure 5.5.1.1.

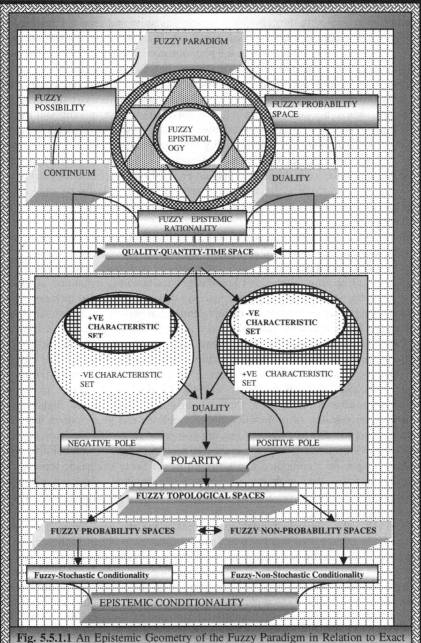

**Fig. 5.5.1.1** An Epistemic Geometry of the Fuzzy Paradigm in Relation to Exact and Inexact Sciences and Reasoning

The structure of the epistemic process is such that in the fuzzy paradigm, we simultaneously deal with the behavior of quality, quantity and time as embodied in an epistemic element in the universal object set. If information is fuzzy, then it may be assumed to be full relative to the phenomenon of interest. The information structure is said to be defective due to fuzziness. In this case, we are placed in the *fuzzy-non-stochastic topological space* where uncertainty is generated solely by the fuzzy process containing ambiguity, vagueness and subjectivity that produce inexactness where the analytical structure involves finding the conditional *exact-value equivalence*. The variable representation of the epistemic elements is *fuzzy-non-stochastic variable*. The conditionality is conceptually expressed through the fuzzy covering of degrees of exactness associated with the phenomenon, and hence we only have to deal with the phenomena and measurements of *fuzzy non-stochastic uncertainty and risk* that are measured in a simple fuzzy topological space and computed with fuzzy mathematics and analyzed with fuzzy logic and its laws of thought.

Alternatively, when we assume incomplete or limited information in the terrain of vagueness, ambiguity and subjectivity, then we are dealing simultaneously with fuzziness and stochasticity in the information-knowledge production. The uncertainty surrounding claims and knowledge production is generated by either *fuzzy-stochastic* or *stochastic-fuzzy process* that produces inexactness. The relevant variable representation of the epistemic elements in the universal object set is either *fuzzy-stochastic variable* in the fuzzy-stochastic space or *stochastic-fuzzy variable* in the stochastic-fuzzy space and the concept of exactness must meet conditions of *stochastic fuzzy conditionality* [R4.48] [R11.22]. The case is such that the analytical structure involves finding conditional *exact-certainty-value equivalence* that allows us to analytically deal with inexact or vague probability measures. It is here that we encounter concepts of higher-level probability, probability of probability, belief functions and many others that still retain the classical laws of thought and are computed in the classical topological spaces. The view in this monograph is that the problem of inexact probability must be defined; specified, computed and analyzed in the fuzzy-stochastic topological space, otherwise the problem of inexact probability acquires phantom characteristics in the sense that is formulated in inappropriate space where solution does not exist.

To develop the concept of *fuzzy conditionality* and the required measure, we must visit the concepts of polarity and duality of the same object in the universal object set. Both polarity and duality are made up of opposite characteristic sets in unity and continuum that admit of internal negation and subjective decision for epistemic identity which ensures a comparison of the epistemic identity with ontological identity of the same phenomenon for knowing. The unity connects the polarity and duality while the continuum is partitioned by fuzzy conditionality that logically allows discrete values to be established within the continuum through decision-choice actions. It is this fuzzy conditionality that provides a logical connectedness between inexactness and exactness of phenomena and then connects the classical paradigm to the fuzzy paradigm. With fuzzy conditionality, the applicable domain of exactness can be specified through the *principle of fuzzy decomposition* where the indeterminacy of vagueness and ambiguity can be moved to determinacy by the fuzzy process and the fuzzy decision-choice rationality in language, reasoning and measurement. The fuzzy decision-choice rationality provides us with logical channels to integrate subjectivity in relation to quality and free the cognitive agent from the tyranny of the classical world of exact rigid determination that is independent of the judgments of thinking beings.

Every exact proposition generated in the classical paradigm has its fuzzy-conditionality covering. The fuzzy conditionality of exactness of any proposition is dependent on the methods of fuzzification and defuzzification. The fuzzification transforms the problems in the classical two-valued exact logical space into the logical continuum in inexact logical space of fuzzy system. The defuzzification transforms the logical continuum in the inexact space of fuzzy systems into qualified exact propositions in the classical two-valued exact logical space. Alternatively stated, dualism with the excluded middle is transformed by fuzzification operator into *inexact-exact duality* with a continuum which is then retransformed by defuzzification operator into dualism with qualification of fuzzy conditionality.

**Definition 1.3.3.1: Fuzzy Conditionality**

Fuzzy conditionality is a conditional expression in a quantitative disposition that specifies the optimal conditions to which a defuzzified variable or statement from a fuzzy topological space belongs to an exact classical topological space with a conceptual or numerical clarity. It specifies the threshold partition of exactness (non-vagueness) from inexactness (vague or ambiguous) as decision-choice determined. The defuzzified value is the measure of the exact-value equivalence with a fuzzy conditionality as a measure of its degree of attached confidence.

The fuzzy paradigm provides us with a vehicle to acknowledge the presence of vagueness, ambiguity, inexactness, qualitative disposition and subjectivity in the information used as input in the knowledge-production process as we move through the quality-quantity-time space. It also offers us a way to develop methods of symbolic representation of defective information structure for reasoning to abstract exact-value equivalences from inexact systems. The solution is such that the classical principle of absolutism is shown as limiting the exact-inexact value cases where any point of separation must meet the conditions of fuzzy conditionality. We may note that series of values of fuzzy conditionality may be used to decompose the fuzzy set of logical truth into its classical parts. The measure of the fuzzy conditionality is the measure of degree of confidence attached to exact and inexact propositions. This allows us to speak of fuzzy propositional calculus in the fuzzy paradigm with different laws of operations and true-false acceptance (for fuzzy propositional calculus see [R3][R3.2][R3.6] [R3.21] [R3.32] [R3.50] [R3.52]).

We may conclude that the examinations of the principal forms of thought of the classical and the fuzzy paradigms and their relative approaches in relating them to the behavior of cognitive agents over the epistemological space constitute the essential forces of the analytical work in the theory of the knowledge square. The epistemological space is shown to follow a universal principle of knowledge production that is revealed by the theory of the knowledge square. The methodological space is partitioned into specificities of knowledge areas and preferences of approach to the knowledge search and the construction of the house of knowledge. The input into the knowledge production is seen in terms of the assumptions made in the use of the defective information structure as an input and the condition of the claim of the knowledge item. At the level of information representation, the classical paradigm is connected with exact symbolism that involves discrete and digital conditions. At the level of information representation, the fuzzy paradigm is connected with inexact symbolism encompassing continuum and analog conditions. Table 5.5.1 is used to show the similarities and differences of the two paradigms.

**Table 5.1.1** Comparison of the Essential Similarities and Differences of the Classical and Fuzzy Paradigms of Thought.

| THE EPISTEMIC PILLARS OF THE CLASSICAL PARADIGM OF THOUGHT | THE EPISTEMIC PILLARS OF THE FUZZY PARADIGM OF THOUGHT |
|---|---|
| 1. Principle of opposites, | 1. Principle of opposites, |
| 2. Dualism, | 2. Duality, |
| 3. Excluded Middle, | 3. Continuum, |
| 4. Mutually exclusive and collectively exhaustive negative and positive characteristic sets, | 4. Mutually non-exclusive and collectively exhaustive negative and positive characteristic sets, |
| 5. Exact Polarity of negative and positive poles, | 5. Inexact Polarity of negative and positive poles, |
| 6. Category:<br>a) Exact-category formation,<br>b) Exact analytical categories,<br>c) Exact nominalism,<br>d) Constant quality,<br>e) Quantity-time space, | 6. Category:<br>a) Inexact-category formation,<br>b) Inexact analytical categories,<br>c) Inexact nominalism,<br>d) Varying quality,<br>e) Quality-quantity-time space, |
| 7. Categorial changes:<br>a) Exact categorial conversion,<br>b) Exact categorial moment,<br>c) exact constructionism,<br>d) Exact reductionism, | 7. Categorial changes:<br>a) Inexact categorial conversion,<br>b) Inexact categorial moment<br>c) Inexact constructionism<br>d) Inexact reductionism |
| 8. Classical Logic and Mathematics:<br>a) Exact Symbolism,<br>b) Exact analysis,<br>c) Absence of subjectivity,<br>d) Exact quantitative disposition,<br>e) Exact rigid determination,<br>f) Exact Probability,<br>g) Exact stochastic conditionality,<br>h) Absence of fuzzy conditionality, | 8. Fuzzy Logic and Mathematics:<br>a) Inexact Symbolism,<br>b) Inexact analysis,<br>c) Presence of subjectivity,<br>d) Inexact quantitative disposition,<br>e) Inexact flexible determination,<br>f) Inexact Probability,<br>g) Inexact stochastic conditionality<br>h) Fuzzy conditionality |
| 9.Exact epistemological Space with defective information structure due to Incompleteness, | 9.Inexact epistemological Space with defective information structure  due to Incompleteness and Fuzziness, |
| 10. Exact decision-choice actions | 10. Inexact decision-choice actions |

The epistemic comparative structure for the essential similarities and differences brings to the understanding how the problem of the classical excluded middle in the classical paradigm is resolved with the fuzzy paradigm with the logical continuum. At the level of thought, the classical paradigm deals with exactness as expressed in dualism without connection to its opposites through the principle of excluded middle. At the level of thought, the fuzzy paradigm deals with inexactness of reasoning as expressed in duality with connection to its opposites through the principle of logical continuum.

# Chapter 6
# Fuzziness, Science, the Knowledge Square and the Problem of Exact Science

Given the structure of the theory of the knowledge square, as presented in previous chapters, we will now reflect on the structure of science and the theory of knowledge in relation to fuzziness that covers ambiguities, vagueness and quality in the information structure as an input for knowledge production. The stage of our modern knowledge, the speed of the global technological progress, the relationship between technological progress and knowledge, the required efficiency of organizational operations to use the human and non-human resources to improve human conditions and maintain life, the required efficiency of social management of synergetic relations, the increasing need for information and the complexities of our accepted knowledge, demand an important reexamination of certain philosophical claims on modes of reasoning and methods of acceptance of epistemic items as true or false in the knowledge-production process. Our knowledge-production enterprise has been partitioned into science and non-science with more social respect and credibility given to the results of science and the enterprise of science. Let us examine the structure and form of this partition.

## 6.1 The Partition of Science into Exact and Inexact Sciences

In the process of knowledge production, the claim of what constitutes scientific knowledge has also been partitioned into *exact* and *inexact* sciences. The implication is that there is a segment of our scientific enterprise that deals with inexact science by some accepted criteria and there is a segment that deals with exact science. This brings us to a conflict between definition and criteria as to what is knowledge, what is science, what is exact science and what is inexact science, and what criteria may be used to demarcate them for knowing and understanding. The establishment of the framework belongs to the problem of demarcation of the knowledge-production space. Knowledge, science, exact science and inexact science epistemologically constitute an awareness of a set of *what there is (ontological elements)* and knowability of *what there is* (epistemological element). We shall present synthetic and analytical structures that all these stated logical entities represent categories of knowing with a category that constitutes a *primary* one from which other categories emerge by *categorial conversion* in logical systems into categorial derivatives.

K.K. Dompere: The Theory of the Knowledge Square, STUDFUZZ 289, pp. 95–115.
springerlink.com                    © Springer-Verlag Berlin Heidelberg 2013

The ontology of these categories requires us to produce the defining characteristics that show the nature of knowledge and then the distinction between science and non-science in the knowledge-production system that will bring into focus the similarities and differences of the exact and inexact sciences, and then show the places they occupy in the knowledge-production process. Our focus then is to answer the question of what are exact and inexact sciences in the knowledge-production process and to what extent are they similar or different. The need for such an epistemic inquiry is driven by our contemporary progress in informatics, biometrics, synergetics, energetics, and intelligent technology in decision-choice sciences, the problems in medical decisions, the structure of social sciences and the knowledge problems of macro-systems. This need has been amplified by the cross-fertilization of different areas in the knowledge-production enterprise in terms of interdisciplinary efforts and activities and other areas of health sciences that have given rise to the complexity theory and the theory of soft or inexact computing. The point of entry into the examinations of the epistemic problems of exact and inexact sciences is from personal encounters with some methodological difficulties in economic science [R2.9] [R11.19], the research works in the general decision theory and a search for epistemic rationality in complex systems that are self-organizing, self-correcting and self-regulating such as macroeconomics, development economics, macroeconomic planning and political economy [R14.30] [R16.1][R16.2][16.4][R16.7]. Discussions in the academic and general literature are riddled with sharp criticisms against economics as a sector of knowledge production. Whenever something goes wrong in our macroeconomic systems, the subject area of economics and its practitioners are singled out for criticism about inexactness, non-scientific status and lack of knowledge in economic thinking.

All the constructed laws of behavior and contributions by economics to different areas of thought regarding the efficient management of society and its institutions, the understanding of uncertainty, the efficient use of resources and the contributions to tools of operations research and the theory of statistical inference among many are all forgotten [R1] [R10] [R11] [R13] except for the notion that economics does not offer us correct policy. The critics forget the Keynes' advice that: *The theory of economics does not furnish a body of settled conclusions immediately applicable to policy. It is a method rather than a doctrine, an apparatus of the mind, a technique of thinking, which helps its possessor to draw correct conclusions* [R 14.30, p.xiii]. From the conditions of knowing, distinctions must be made among correctness, precision, accuracy and exactness epistemic claims. This statement of Keynes may be extended to other areas of knowledge. Again, an important point is always overlooked. This point is simply that the practices and the uses of economic methods and the epistemic frame of economics, as apparatuses of the mind and techniques of reasoning and a framework to create economic policies, are influenced not by the validity of economic thought but by the political expediencies and the ideological positions of the ruling political governance. In the practice of the management of the social systems, political rationality comes to replace economic rationality such that good economic policies may be sacrificed on the basis of political rationality. On these bases, economics, for example, is placed in the category of inexact science. What attributes constitute the characteristics of science are presented through assumptions rather than critical *explication* of the central building blocks of the knowledge production and the basic explication of the key defining terms such as science and knowledge itself.

## 6.2 Problems of Explication and Definition in the Knowledge-Production Process

In general, explication involves the linguistic process of cognitively transforming an inexact ordinary term (*explicandum-* that which is to be explicated) into a more exact term (*explicatum-* that which explicates) that is termed scientific in its linguistic home [R2.9] [R9.13] [R10.4] [R10.12][R14.18][14.84]. The ordinary linguistic terms, with their rules of constructive relations to define a language, are viewed as vague. Each term contains multiplicity of meaning that widens the applicable space of its uses. The process of explication is to strip the word or symbol of an ordinary language off or reduce its vagueness to a minimum and assign relatively an exact linguistic meaning within a special context in cognition and language. In other words, we work from inexactness to exactness. The concept of special context is very important in the sense that the context helps to define the term's domain of applicability. For example, the term equality must be seen in a context-specific; otherwise, it loses its useful scientific meaning. Equality under law in a society is different from equality in quantitatively computable systems as well as equality in income distribution in economic systems. Similarly, iso-equality curves have specific meaning depending on the context of the area of knowledge. The maps of iso-equality curves in the utility spaces, in output spaces, temperature spaces and magnetic fields present different conceptual system of information representations with differential exact and inexact combinations. The objective of explication is to reduce the dimensionality of the applicable meaning associated with a word or symbol which will require subjective interpretation that may vary across many individual cognitive agents and knowledge areas.

The epistemic framework of demands for explication is that when any problem of the knowledge process is stated in inexact terms, the problem acquires characteristics of inexactness which then prevents an attainment of an *exact solution* without an epistemic conditionality. In the analytical structure of exact knowledge, we must, thus, solve the initial two problems of explication of terms and the optimal process of exact knowledge production. In the process of constructing the knowledge structure, it is useful to keep in mind that the explication problem is different from the knowledge-production problem. There is no satisfactory solution to the knowledge-production problem when there is the presence of indeterminacy of meaning since such indeterminacy may lead to either *phantom* or *ill-posed problems* The solution to the knowledge-production problem cannot be determined to belong to the set of right or wrong solutions or as applied to propositional calculus because the true-false value is not determinable. All these pertain to the concept of exact science in terms of explication of language and terms of scientific communication and for the reporting of the results that must be transmitted to the general knowledge community. It is here that the problems of explications and definitions tend to arise in the knowledge-production process. It is also here that methodological discreteness and methodological continuum reveal their epistemic similarity and differences.

## 6.2.1  Concepts of Definition, Explication and Exactness in Information-Knowledge Representations

The question that always emerges is: what set of required conditions will constitute a satisfactory solution to the problem of explication that will help to specify the domain of exact knowledge, general science and exact science? The problem belongs to theories of meaning and formal languages. The problem also presents itself as, given the *explicandum*, what is the satisfactory *explicatum*? This problem, in an ordinary language, boils down to the problem of hedging. The problem may be stated simply as given an *inexact concept* of a term, what is the acceptable *exact concept* equivalence or term in a specific context of the knowledge production in a given language? In this epistemic process, we must explicate the concepts of inexactness and exactness themselves in relation to the context of the knowledge production. The problem of *explicandum* ($\mathcal{E}d$) and *explicatum* ($\mathcal{E}t$) must also be seen in the context of *inexact-exact duality* with conflicts of meanings and dynamics of categorial conversion in a given language where the meaning of exactness is relatively connected to that of inexactness in a continuum. Let us turn our attention to a set of workable definitions relevant for our discussions on the concepts of inexactness and exactness from which we shall attach science in the knowledge-production process.

### Definition 6.2.1.1a: Linguistic Approach for Inexactness/ Exactness

Inexactness in human communication and reasoning in a given language involves situations where there is a multiplicity of meaning that may be assigned to a term and conclusion depending on the sending and receiving cognitive agents. Exactness is such that one meaning is assigned to the term irrespective of the sender or receiver of the message.

### Note 6.2.1.1

This property is translated to discrete and digital and projects a property of a singleton set. It is this property of the singleton set that makes digital process appealing at the expense of comprehensiveness and clarity of complexity.

### Definition 6.2.1.1b: Symbolic Approach for Inexact and Exactness

Let $\mathbf{T}$ be a term and $\mho$ be the space of meaning in a language $\mathbf{L}$, and let $\mathcal{f}$ be a function defined on $\mathbf{T}$ into $\mho$ to assign meaning to $\mathbf{T}$ in language $\mathbf{L}$, and hence the meaning of $\mathbf{T}$ may be written as $\mathcal{M} = \left\{ \mathfrak{m} \mid \mathbf{T} = \mathcal{f}(\mathfrak{m}), \mathfrak{m} \in \mho \right\}$ where $\#\mathcal{M} \geq 1$.

The term $\mathbb{T}$ is said to be inexact if $\#\mathfrak{M} > 1$ and it is said to be exact if $\#\mathfrak{M} = 1$. It is said to be indefinable if $\mathfrak{M} = \varnothing$ where in this respect $\mathbb{T}$ is said to be the basic or a linguistic primitive. The set $\mathfrak{M}$ is said to be linguistically a *definitional set* for the term $\mathbb{T}$.

Let us now turn our attention to the linguistic and analytical specification of explication. This requires that we first provide the specification of what a definition connotes and how the concept of a definition relates to that of an explication. Definition is a relationship between that which is to be defined called *definiendum* and that which defines called *definiens*. Definition, therefore, is a reasoning device or a process to fix the meaning of a symbolic representation in a given language $\mathbb{L}$, or by the use of an extension principle to enlarge the language $\mathbb{L}$, with new symbolic representations over the epistemological space. As stated, every element $\mathfrak{m} \in \mathfrak{M}$ is a definition of a term $\mathbb{T} \in \mathbb{L}$. Each of the defining terms in the set $\mathfrak{m} \in \mathfrak{M}$ may be viewed in terms of the degrees of exactness to which the meaning is a representation of $\mathbb{T} \in \mathbb{L}$. If $\mathfrak{m}_1, \mathfrak{m}_2 \in \mathfrak{M}$, then $\mathfrak{m}_1$ and $\mathfrak{m}_2$ are in equivalence relation $(\approx)$ in definability if they have the same degree of exactness of definition, and hence we write $\mathfrak{m}_1 \approx \mathfrak{m}_2$ as interchangeable in the language $\mathbb{L}$. It is the ranking of meanings of words by the distribution of the degrees of exactness of meaning, given a linguistic term, which provides the efficiency of language in relation to thought and expressions of ideas.

## 6.2.2 Exactness, Fuzzy Definitional Set and the Construct of the Explicator Set

The set $\mathfrak{M}$ with index set $\mathfrak{I}$ is a fuzzy set of meanings in degrees of exactness relative to $\mathbb{T} \in \mathbb{L}$. As a fuzzy set of meanings, it may be represented as $\mathfrak{M}$ with membership characteristic function specified as $\mu_{\mathfrak{M}}(\mathfrak{m}) \in [0,1]$ where the degree of closeness to which $\mathfrak{m} \in \mathfrak{M}$ defines $\mathbb{T} \in \mathbb{L}$ is ranked by the degree of belonging $\mu_{\mathfrak{M}}(\mathfrak{m})$, thus, $\mathfrak{m}_1 \approx \mathfrak{m}_2$ if $\mu_{\mathfrak{M}}(\mathfrak{m}_1) = \mu_{\mathfrak{M}}(\mathfrak{m}_2)$. Furthermore, given $\mathfrak{m}, \mathfrak{m}_2, \mathfrak{m}_3 \in \mathfrak{M}$ if $\mu_{\mathfrak{M}}(\mathfrak{m}_1) > \mu_{\mathfrak{M}}(\mathfrak{m}_2) > \mu_{\mathfrak{M}}(\mathfrak{m}_3)$, then we conclude that $\mathfrak{m}_1$ is a better definitional word among the three in the language $\mathbb{L}$. Let us keep in mind that every language, $\mathbb{L}$ may be divided into *linguistic primitives*, the indefinable words, $\mathbb{L}_\pi$ and linguistic derivative $\mathbb{L}_\delta$, with the definable words and a grammar $\mathbb{L}_\gamma$ where $\mathbb{L} = \mathbb{L}_\pi \cup \mathbb{L}_\delta \cup \mathbb{L}_\gamma$. The set of definable terms creates linguistic conditions of substitutions and transformations in communication and thought processes conditional on $\mathbb{L}_\gamma$. In this respect, a definability condition requires that there exist at least one element, $\mathfrak{m} \in \mathfrak{M}$ such that $\mathbb{T} = \mathfrak{f}(\mathfrak{m})$, and $\mathbb{T}, \mathfrak{m} \in \mho$. The concept of definition in the knowledge-production process requires a set-to-point mapping where if $\mu_{\mathfrak{M}}(\mathfrak{m}_1) > \mu_{\mathfrak{M}}(\mathfrak{m}_2)$ and

$\mu_{\mathfrak{M}}(\mathfrak{m}_2) > \mu_{\mathfrak{M}}(\mathfrak{m}_3)$ then $\mu_{\mathfrak{M}}(\mathfrak{m}_1) > \mu_{\mathfrak{M}}(\mathfrak{m}_3)$ which simply implies consistency of meaning in the context of the subject. Every non-primitive term has a definitional dimension in sciences and the knowledge-production and communicational processes.

The definition of an exact meaning presents some problems since every, $\mathfrak{m} \in \mathfrak{M}$ is also a set of meanings if it is not a linguistic primitive. Given the definability conditions, explication involves a reduction in dimensionality of definitional elements contained in $\mathfrak{M}$ into a new fuzzy set $\tilde{\mathfrak{M}}$ of meanings with an index set $\tilde{\mathfrak{I}}$ and a membership characteristic function $\mu_{\tilde{\mathfrak{M}}}(\mathfrak{m}) \in [0,1]$. Explication, in this respect, is the same thing as a definition with a reduced dimensionality of the form $\underset{\mathfrak{m} \in \mathfrak{M}}{\vee} \mu_{\mathfrak{M}}(\mathfrak{m}) \in [0,1]$, $(\vee = \max)$. The new set formed by a mapping from the definitional set into explicative space is called the *explicator set*. In this respect, the condition for explicability is such that the explicator set is contained in the definitional set. The condition of exactness in explication, given the explicator set, requires, that $\mathfrak{m} = \mathbf{T} \in \mathbf{L}$ in addition to the condition that $\mathfrak{M}$ is a singleton set with $\mu_{\mathfrak{M}}(\mathfrak{m}) = 1 = \mu_{\mathfrak{M}}(\mathbf{T})$. In which case, we have equality between *explicandum* and *explicatum* $(\mathcal{Et} = \mathcal{Ed})$ and hence explication is not needed in the sense of reducing the dimensionality of the definitional set.

Inexactness may be due to vagueness in the meaning of the linguistic term or ambiguities in the instruments of reason or limited information for logical derivatives. Both inexactness and exactness are categories of meaning and reasoning. In human linguistic structures and from the viewpoint of explication, the category of inexactness in meaning and reason is taken as the primary category for logical transformations. Exactness is a derived category from the primary category of inexactness where both exist as duality in logical unity requiring a categorial conversion in continuum. In other words, the search for exactness in the knowledge production is arrived at by a process from inexactness but not the other way around. Exactness, therefore, cannot be assumed as an epistemic characteristic of any language and in the domain of cognition. Here, we distinguish between ontological exactness and epistemological exactness. Inexactness is not a characteristic in the ontological space. In the epistemological space, however, the activities of knowledge production are such that we begin with an inexact representation and work our way to exact representation in all areas of cognition. This is equivalent to the statement that from nothingness we search for *somethingness* in the continuum of *nothingness-somethingness* duality or in the ignorance-knowledge duality in a continuum. The concepts of exactness and inexactness are only meaningful in the epistemological space where we seek to establish the identity of epistemic elements with the ontological elements.

Inexactness is due to information incompleteness (partial ignorance) and information vagueness (ambiguities or lack of clarity). Thus the solution to the problem of explication requires us to solve two problems of *information completeness* (incompleteness) and *information clarity* (vagueness). Both problems cannot be solved to provide full exactness of concepts in moving from *explicandum* (concept of inexactness) to *explicatum* (concept of exactness). The problem is not futile when we realize that inexactness and exactness in the epistemological space are in linguistic degrees of preferences in meaning the cutoffs of which are decision-choice determined. This decision-choice determination allows us to accept, on the practical level, that the needed concept has been massaged enough to be, at least, practically exact to serve as a basis of knowledge production. In this case, the value of $\mu_{\hat{\mathfrak{M}}}(\mathfrak{m}) \in [0,1]$ has been determined. Explication, therefore, is a subjective maximization of exactness of meaning under constraint of inexactness of meaning so as to reduce the dimensionality of the definitional set of meanings. Analytically, it is an optimization of the membership characteristic function of exactness (inexactness) subject to the membership characteristic function of inexactness (exactness). For further analysis of explication see [R10.4][R10.12]R14.18][14.84] [R19.52]. Exactness, therefore, is an ultimate state of perfect meaning toward which cognitive process may seek as a goal. The nature of human information-decision-choice process is such that there will always be an *irreducible core of inexactness* that is generated by defective information structure and ambiguities in reasoning to preserve the irreducible ignorance.

**Definition 6.2.2.1: Irreducible Core of Inexactness**

An irreducible core of inexactness is a notion where the definitional set cannot be massaged to reduce its dimensionality for the creation of the explicator set. The measure of the irreducible core of inexactness is the value of the *fuzzy residual*.

The fuzzy residual may also be seen in terms of irreducible core of vagueness. In the epistemological space, we observe the presence of exactness-inexactness duality in the definitional set $\mathfrak{M}$ with a general characteristic set, $C$, for each element $\mathfrak{m} \in \mathfrak{M}$ of the characteristic set of exactness, $E$, with a generic element, $e \in E$ and a membership function of $\mu_E(e)$. Alongside of it we have the characteristic set of inexactness, $Z$, with a generic element of $z \in Z$ and a membership characteristic function, $\mu_Z(z)$ where , $C = (E \cup Z)$ is the general characteristic set such that $e, z \in C$. The exactness-inexactness duality exists in a continuum such that any inexactness has exactness support and vice versa. We shall refer to this inseparable existence as *fuzzy logical support*. The structures of the characteristic functions are such that $\mu_E(e)$ is upward sloping ($\left( \dfrac{d\mu_E(e)}{de} \right) \geq 0$) while $\mu_Z(z)$ is downward slopping $\left( \dfrac{d\mu_Z(z)}{dz} \right) \leq 0$ with $e, z \in C = (E \cup Z)$. The curvatures of the membership functions project the notion that as the characteristics of inexactness get reduced, the characteristics of exactness increase where there is a point of indifference. They combine to project exact-inexact duality with fuzzy continuum. The explication problem is then solved by optimizing the membership function of the exactness characteristic set subject to the membership function of the characteristic set of inexactness. The process requires the framing of a fuzzy decision problem $D = (E \cap Z)$ with a membership function defined as $\mu_D(\cdot) = [\mu_E(\cdot) \wedge \mu_Z(\cdot)]$ the optimum of which will then be sought. The solution to the decision problem may be formulated as a fuzzy optimization one of the form:

$$\alpha^* = \mu_D(e^*) = \underset{e}{opt}\, \mu_D(e) = \begin{cases} \underset{e \in D}{Opt}\, \mu_E(e) \\ s.t \left[ \mu_E(e) - \mu_Z(e) \right] \le 0 \end{cases} \qquad (6.2.2.1)$$

Each element in the definitional set will have an associated optimal $\alpha^*$ which will then be used to construct the *explicator set*, $\aleph$ through the method of fuzzy decomposition with

$$\aleph = \left\{ m \mid \mu_{\mathcal{A}}(m) \ge \alpha^* \in (0.5, 1) \right\} \qquad (6.2.2.2)$$

**Definition 6.2.2.2: Explicator Set**

An explicator set is the collection of definiens with an associated distribution of the degrees of definitional exactness that are greater than the indifference point. An explication is said to exist if there is an explicator set that can reduce the dimensionality of the definitional set of a word or term in concepts.

It is generally held that given an explicandum (word of ordinary language which is to be explicated), the explicatum (word of science that explicates) must satisfy the conditions of explicandum. Similarly, the exactness in a word's application, the usefulness in its information carriage and the concept's simplicity to convey meaning must satisfy the conditions of the explicandum. The process of explication, therefore, is to establish the language of knowledge and define the grammar of science where there are many grammars as there are areas claimed to be science. The condition where the explicatum is equal to explicandum $(\mathcal{E}t = \mathcal{E}d)$ is difficult to meet. This difficulty is resolved by the use of decision-choice action, in that, exactness is determined by cognitive resolutions in the conflicts of meanings in any language as we have established in eqns. 6.2.2.1 and 6.2.2.2. It is only *conceptually neutral* mathematical and symbolic logic that exactness can be achieved where every symbol in the mathematical language meets the condition of reflexivity where $(x = x)$ and $(y = y)$ and if $(x = y)$ then they represent the same thing when concepts are attached. These conceptually neutral mathematical and logical symbols escape the problem of inexactness in the sense that $x$ is $x$, with membership degree of exactness equals to one, and with zero irreducible core of inexactness such that it can be operated on with exact laws of thought. In this case, the *law of identity* is met in the sense that x is what x is, and hence assumes the characteristics of an ontological element where the conditions of the ontological identity are assumed to hold in the epistemological space (see the discussions in [R9.13] [R14.3][R14.5] [R14.6] [R14.14] [R14.15] [R14.31] [R14.68] [R14.113] [R19.4][R19.58]).

The decision-choice approach to solve the explication problem, like any decision-choice element in human action, involves two types of uncertainties. One type of the uncertainty is due to *information incompleteness* and the other type is due to *information vagueness*. The information incompleteness provides us with *stochastic uncertainty* and *probabilistic belief structures*. Probability is said to exist if there is defective information structure due to quantitative incompleteness of the available information. The information vagueness provides us with *fuzzy uncertainty* with *possibilistic belief structures* about our acceptance of the derived explicatum. Thus a possibility is said to exist if there is a defective information structure due to qualitative information vagueness and ambiguity. This decision-choice process must lead to a classificatory system of concepts about elements in the universal object set that will allow us to define distinctions and similarities in terms of qualitative and quantitative comparisons for logical category formations that are relevant to concept representations, thought formations and validity analyses with logical conversions from ignorance or partial ignorance to knowledge, and from unjustified opinions to justified opinions.

## 6.3   Ontological Space, Epistemologial Space and Defective Information Structure

In all these, we cannot avoid the notion that the results of knowledge are inseparable from human cognitive actions involving information-decision-interactive processes. Here again and as elsewhere, two types of information are distinguished [**R2.9**]. One type is objective information which is associated with ontological elements and the other type is subjective information which is associated with epistemological elements. In this way, ignorance is defined as *epistemic distance* between ontological and epistemological elements for any given element in the universal object set. The information incompleteness and information vagueness are defined in the subjective information space which is under the control of cognitive agents. The concepts of vagueness, ambiguity, incompleteness, and subjectivity are meaningless and have no epistemic contents in the world devoid of cognitive agents. In these discussions, objectivity is to ontology just as subjectivity is to epistemology. Over the ontological space objectivity is seen in terms of identity without subjectivity. Over the epistemological space objectivity is seen in terms of subjective set covering defined as circular distance around the ontological identity.

### 6.3.1   Defective Information Structure and Cognition

Ignorance is reflected in the lack of knowledge and the lack of knowledge is reflected in the deficiency in the composite information and thought involving cognitive activities over the epistemological space. The composite aggregate of information incompleteness and information vagueness constitutes the *defective information structure*.

**Definition 6.3.1.1: Defective Information Structure**

A defective information structure is an epistemological phenomenon in the process of reducing human ignorance by cognizing the ontological elements through epistemic processes. The information structure is said to be defective if the characteristic set of the epistemic elements are incomplete and vague.

**Note 6.3.1.1**

If the characteristic set of an epistemic element is incomplete then the information structure is said to be quantitatively defective. It is said to be qualitatively defective if some elements of the characteristic set of the epistemic element are vague. The information structure is said to be partially defective if it is either qualitatively or quantitatively defective but not both. It is said to be completely defective if it is both qualitatively and quantitatively defective. The qualitative and quantitative defectiveness may be specified in terms of membership characteristic functions of set creation to deal with qualitatively and quantitatively linguistic variables.

The information structure defining ontological elements is not under the control of cognitive agents as such it is taken to be objective since the ontological elements are what they are. The information structure, defining the epistemological elements, is under the control of cognitive agents as such it is taken to be subjective since the epistemological elements are what we claim to be under cognition. The relationships among ontology, epistemology and information may be illustrated with Figure 6.3.1.1.

The explication of exact science, then, boils down to answering a question of whether there exists any knowledge (epistemic) item that is of ultimate certainty that defiles all cognitive doubts and vagueness given any program of human cognition or an epistemic process with epistemological information under either discreteness or continuum. The ontological question of existence of an ultimate certainty of knowledge and hence a property of exactness of an epistemic item goes directly into the nature of human conception and the required toolbox of knowledge search, information processing and knowledge acquisition. This question must be related to the epistemological question of how can cognitive agents know the ultimate certainty of knowledge of any ontological item. Hopefully, the ontological problem of existence of ultimate certainty of knowledge and hence exact science is not a phantom problem in the sense of Max Planck [R8.55]. To make sure that we are not dealing with a phantom problem, we may define an ultimate certainty of knowledge.

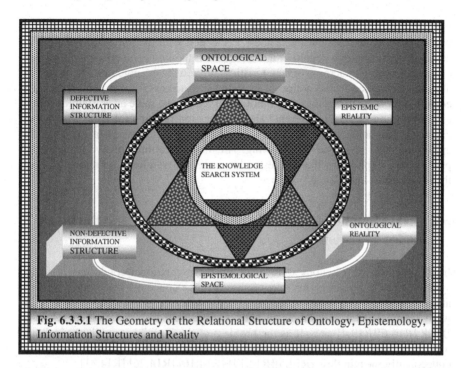

**Fig. 6.3.3.1** The Geometry of the Relational Structure of Ontology, Epistemology, Information Structures and Reality

## Definition 6.3.1.2: Ultimate Certainty of Knowledge

An epistemic item is said to contain an ultimate certainty of knowledge if its epistemic characteristic set can be shown to be equal to the ontological characteristic set of the ontological element such that $\mathbb{X}_{C_e} = \mathbb{X}_{C_\mathfrak{A}}$ where $\mathbb{X}_{C_e} \subseteq \mathbb{X}_{C_\mathfrak{A}} \subseteq \mathbb{X}_{C_e}$.

## Note 6.3.1.1

The above definition must be referenced to the Note 5.3.1 of Chapter 5 where Ontological category is $\mathbb{C}_\mathfrak{A}$, Ontological characteristic set is $\mathbb{X}_\mathfrak{A}$, Epistemological category is $\mathbb{C}_e$, and Epistemological characteristic set is $\mathbb{X}_e$. In the ontological space, the characteristic set of each category may be identified as $\mathbb{X}_{C_\mathfrak{A}}$ while in the epistemological space, the characteristic set of each category may be identified as $\mathbb{X}_{C_e}$. The problem of ultimate certainty of knowledge is said to a phantom one if it does not exist in an ontological sense and hence. $\mathbb{X}_{C_e} \subset \mathbb{X}_{C_\mathfrak{A}}$.

To examine the ontological problem, we must answer a number of epistemological questions. Are the principles of cognition and toolbox of exact science created on any universally valid principle? If they are, what is the universally valid principle for their support? How are truth and falsity established with this universally valid principle and how are cognitive elements admitted into the universal knowledge bag or into the epistemic reality? Can we find a world view that is generally and uniformly accepted by knowledge finders? Are the properties of exactness and inexactness associated with the *universal ontology* or associated with human cognition and hence with the *universal epistemology*? Are exactness and inexactness characteristics of actual or potential or both? Is absolutism implied in the notion of knowledge exactness? Is epistemic exactness equal to ontological exactness? Is ontological reality the same as epistemic reality, and if not, how are they different? Here, a question is being raised as to whether exactness, vagueness and inexactness are ontological properties of existence or epistemological properties of knowing. All these questions involve a cluster of problems of explication that relates to language and symbolic representations of the information-knowledge system.

## 6.4    Exact Symbolism, Intuitionist Mathematics and the Fixed Point Theorem

In chapter Five, discussions were made on representational conditions of fuzzy symbols or other symbols that may represent vague concepts. The vague concepts are represented by the word of the entity and the degree of exactness to which the word (symbol) represents the concept's entity. This was the problem of Max Black [R14.6][R19.4] as well as the problem of the intuitionist mathematical and logical school that rejects the formalist exact symbolic representation as conceptually meaningless [R14.5][R14.14] [R14.16] [R14.26] [R14.31].

### 6.4.1    Qualitative Disposition, Vagueness and Exact Symbolism

When inexactness or vagueness is associated with quality, it complicates the concept's symbolic representation as well as introduces subjective phenomena into the thought process in that both quality and quantity must be simultaneously represented by a symbol in both static and dynamic systems. The symbolic representations of concepts and thoughts acquire extra complexities as the qualitative disposition is considered alongside of qualitative disposition with the passage of time. This case was of a particular concern for Karl Niebyl [R14.82]. In this connection on information-knowledge representation and the challenges that it presents in thought production and the principles of symbolism, it is useful to reflect on the statement by Russell.

> Reflection on philosophical problems has convinced me that a much larger number than I used to think, or than is generally thought, are connected with the principle of symbolism, that is to say, with the relation between what means and what is meant. In dealing with highly abstract matters it is much easier to grasp the symbols (usually words) than it is to grasp what they stand for. The result of this is that almost all thinking that purports to be philosophical or logical consists in attributing to the world the properties of language. Since language really

*occurs, it obviously has all the properties common to all occurrences, and to that extent the metaphysic based upon linguistic considerations may not be erroneous. But language has many properties which are not shared by things in general, and when these properties intrude into our metaphysic it becomes altogether misleading. I do not think that the study of the principles of symbolism will yield any positive results in metaphysics, but I do think it will yield a great many negative results by enabling us to avoid fallacious inferences from symbols to things. The influence of symbolism on philosophy is mainly unconscious; if it were conscious it would do less harm. By studying the principles of symbolism we can learn not to be unconsciously influenced by language, and in this way can escape a host of erroneous notions* [R19.58, p. 147]

Vagueness in symbolic representation of concepts and the reasoning with vague concepts present important challenges to the classical paradigm, with its logic and mathematics, that projects exact rigid determination under the principles of excluded middle and non-contradiction. These Aristotelian principles of excluded middle and non-contradiction are rejected by intuitionist logic and mathematics. The problem is that we have to solve the problem of simultaneous representation of quality and quantity with neutrality of time. One reason is that when logical and mathematical symbols become devoid from contents, they loose the characteristics of qualitative disposition of epistemic elements. In this way, their analytical manipulations and conclusions that are derived carry no identifiable knowledge, and the derived truths or falsities are only internally valid within the system's axiomatic reasoning process. In this respect, reasoning with symbolic exactness produces only a series of relations that is independent of content and meaning that one may want to assign to the relations of the entities that are produced by reasoning. The epistemic framework of the classical exact rigid determination is constructed on *the presupposition of existence of a world of mathematical objects, a world independent of the thinking individual, obeying the laws of classical logic and whose objects may possess with respect to each other the 'relation of a set to its elements* [R14.14 p.81]. This is supported by the statement: *The law of excluded middle is true when precise symbols are employed, but it is not true when symbols are vague, as, in fact, all symbols are* [R19.58, p.149]. A further supporting statement of the above is: *All traditional logic habitually assumes that precise symbols are being employed. It is therefore not applicable to this terrestrial life, but only to an imagined celestial existence* [R19.58, p151].

The foundations of the development of paradigms of thought begin with the nature of information structure and its symbolic representation for the knowledge production over the epistemological space. The nature of the principle of symbolism in the information representation will dictate the type of reasoning that will be appropriate in symbolic manipulations and thought production. Since language, for expressing information, carries with it vagueness and the available information as an input of reasoning is limited, it is useful to begin with the notion that the information for initializing the primary category in constructing the derived categories of knowledge over the epistemological space  is defective. Thus, the problem, involving the analytical work of reasoning with symbolic representation of concepts and their relationships in any given language, is that we have to deal with a defective information structure that is due to vagueness and incompleteness in information signals in the transmission-reception processes. The interactive relationships of the methodology of representation and analysis may be viewed in terms of exactness, discreteness, inexactness and continuum as in Figure 6.4.1.1, where one pyramid presents an exact methodological path that is associated with the classical paradigm of exact rigid determination while superimposed on it is another pyramid that presents a methodological path of inexactness with fuzzy paradigm of inexact flexible determination. The pyramidal structure of exactness-discrete-methodological space represents the path of the classical paradigm in the enterprise of the knowledge production. The pyramidal structure of inexactness-continuum-methodological space represents the path of the fuzzy paradigm in the enterprise of the knowledge production.

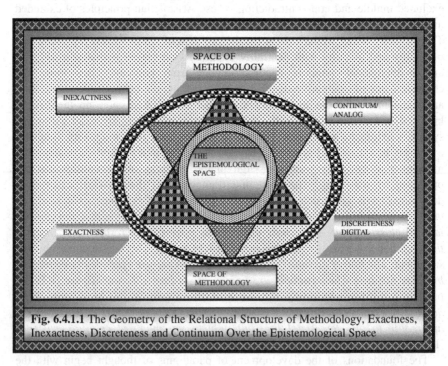

**Fig. 6.4.1.1** The Geometry of the Relational Structure of Methodology, Exactness, Inexactness, Discreteness and Continuum Over the Epistemological Space

The questions then become: How do we deal with the problem of vagueness in symbolic representation of vague concepts and ideas, and what reasoning rules must be created if we reject some principles of the classical paradigm over the epistemological space? These questions and the answers have been points of critical contentions and lines of separation between the formalists and intuitionalists schools of mathematics and logics [R14.3] [14.5][R14.14]] [R14.96] [R14.99]. They are also the points of difference and the lines of separation between the *classical paradigm* and the *fuzzy paradigm*. Given the acceptance of information incompleteness, we must have a way of constructing symbols for information-knowledge representation under conditions of vagueness and ambiguities that are characteristics of language and reasoning. This is the problem of vagueness in representation and thought. The problem also relates to exactness and inexactness, discreteness and continuum and digital and analog which belong to deformity of the epistemological space.

## 6.4.2 Approaches to Solving the Vagueness Problem in Symbolism

Two different paths are opened to us for solving the vagueness problem as we view the foundation and evaluation of symbolism in logic and mathematics in relation to the positions of the formalists and the intuitionalists. One path follows a process to conceptually get rid of the qualitative properties of vagueness and ambiguities associated with ordinary language and reasoning, by imposing an assumption of exactness on the information-knowledge structure or through working with the micro-units of conceptual objects in digits that in turn allows exact symbolism. If one works with the micro units, then the second sequential order problem is to find the conditions that will allow the construct of exact symbolic representation of the concepts and thought formation. This is the path taken for the development of the classical paradigm with its logic and mathematics involving discreteness and digital values [R14.96] [R14.4] [R14.6] [R14.5] [R8.5]. This is shown in Figure 6.4.1.1 as the methodological exactness where the fundamental discreteness is claimed by doing away with qualitative aspects of defective information structure.

The second path follows a process of accepting the qualitative value of vagueness and ambiguities associated with the ordinary language and cognition with regard to the defective information-knowledge structure over the epistemological space where decision-choice actions and subjective judgments are the driving force of the knowledge-production process. This is the path taken for the development of the fuzzy paradigm with its logic and mathematics [R3] [R3.54] [R3.57] [R4] [R4.117] [R4.118] [R5] [R5.16][R5.32] [R6.26]], and also the intuitionalist paradigm with its logic and mathematics that involve analog, continuum and interval mathematics where every logical value is a set whose boundaries are fixed by a decision-choice rationality with a subjective judgment under fuzzy conditionality [R14.5][R14.15] [R14.31][RI4.3].

Since much of the discussions in this monograph has been devoted to establishing the epistemic foundations of fuzzy paradigm in relation to the defective information structure, let us turn our attention to the analysis of the conditions of the foundational justification of the classical paradigm with its mathematics and logic in relation to non-defective information structure that will allow the use of exact symbolism. The question is what set of conditions must be met for the use of exact symbolism. Our main concern now is to relate exact symbolism to definitional set, definitional function, explicator set and the fixed point theorem in a given language. The objective is to find out the conditions on the basis of which exact symbolism can be sustained. The fixed point theorem of interest is that of Brouwer [R.11.11]. We have already presented the definitional and explicator sets in the fuzzy system under the conditions of defective information structure over the epistemological space.

The claim is that every knowledge sector must deal with defective information structure that relates to qualitative and quantitative dispositions of all epistemic objects with the passage of time. We shall work with these notions. It is a conjecture here that the motivation of Brouwer's fixed point theorem may have been derived from his involvement in the intuitionalist-formalist debate on exactness in symbolism and the construct of mathematics. We shall show that under the very general conditions, the Brouwer fixed point theorem, as applied to linguistic definitions, denies the existence of exact symbolism in the epistemological space, in that the conditions of the fixed point in any language is so stringent that no language can satisfy it in information-knowledge representations, except when words or symbols are taken as linguistic primitives [R11.11] [R13.2] [R13.13]] [R14.14].

We have stated the relational conditions for definiendum (that which is to be defined) and definiens (that which defines). Consider the discussions in Section 6.2 and definitions 6.2.1.1 (a and b). Now consider the definitional set, $\mathfrak{M}$, and a continuous mapping $\mathfrak{f} : \mathfrak{M} \rightarrow \mathbf{T}$ in a language, $\mathbf{L}$ and the space of definitions $\mathfrak{V}$. The set $\mathfrak{M}$ is linguistically closed, convex and bounded for the term $\mathbf{T}$ in the space $\mathfrak{V}$. The linguistic closure means that there are limited number of words in the language, $\mathbf{L}$, that can serve as definiens for the term $\mathbf{T}$, in the space of meanings $\mathfrak{V}$. The linguistic boundedness means that there is a best definiens that serves as an upper bound and a worst definiens that serves as the lower bound for the definiendum, $\mathbf{T}$, in the space $\mathfrak{V}$ of language $\mathbf{L}$. Linguistic convexity implies that a linear combination of two definienses also belongs to the definitional set $\mathfrak{M}$. The definitional function $\mathfrak{f}(\cdot)$ is a set to point mapping with a set of fuzzy conditionalities. The conditions of any language bring into the analytical structure, crisp and fuzzy definitions and explications that may be presented in terms of relational pyramids of definition and explication.

### Definition 6.4.2.1: Crisp Definitional Set

If $\mu_{\mathfrak{M}}(\mathfrak{m})$ is the distribution of degree of exactness of $\mathfrak{m} \in \mathfrak{M}$ with respect to the meaning of $\mathbf{T} \in \mathfrak{V}$ in the language $\mathbf{L}$, then $\mathfrak{M}$ is said to be a crisp definitional set if $\mu_{\mathfrak{M}}(\mathfrak{m}) = 1$ for all $\mathfrak{m} \in \mathfrak{M}$. It is said to be a fuzzy definitional set if $\mu_{\mathfrak{M}}(\mathfrak{m}) \in (0,1]$ where $\mu_{\mathfrak{M}}(\mathfrak{m})$ is a membership characteristic function from the definitional set $\mathfrak{M}$ into $[0,1]$.

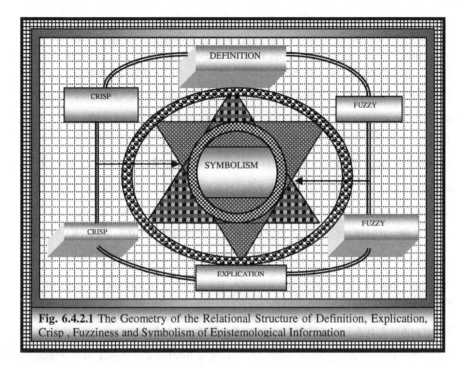

**Fig. 6.4.2.1** The Geometry of the Relational Structure of Definition, Explication, Crisp , Fuzziness and Symbolism of Epistemological Information

**Note 6.4.2.1**

The crispness of the definitional set implies that for all $m \in \mathcal{M}$, $\mathcal{F}(m) \equiv \mathbf{T} \in \mathbf{L}$ with $\mu_{\mathcal{M}}(m) = 1$, $\forall m \in \mathcal{M}$. In this case, any of the definiens meets the conditions of substitutability, not only for the definiendum, but also among each other.

**Lamma 6.4.2.1**

If $\mathcal{M}$ is a fuzzy definitional set for a term $\mathbf{T} \in \mho$ in a language $\mathbf{L}$, then $\mathcal{M}$ is said to be bounded from above if $\exists \alpha \in [0,1]$ such that $\mu_{\mathcal{M}}(m) \le \alpha$ for all $m \in \mathcal{M}$. It is said to be bounded from below if $\exists \beta \in (0,1]$ such that $\mu_{\mathcal{M}}(m) \ge \beta \ne 0$, $\forall m \in \mathcal{M}$ and it is said to be bounded if $\forall m \in \mathcal{M}, \mu_{\mathcal{M}}(m) \in [\beta, \alpha]$, and $\beta, \alpha \in [0,1]$. The $m$ with a value $\mu_{\mathcal{M}}(m) = \alpha$ is called the *maximal exact definiens* while the one with the value $\mu_{\mathcal{M}}(m) = \beta$ is called the *minimal exact definiens*.

The fuzziness of the definitional set implies that for all $m \in \mathcal{M}$, $\mathcal{F}(m) = \mathbf{T} \in \mathbf{L}$ with $\mu_{\mathcal{M}}(m) \in (0,1]$, $\forall m \in \mathcal{M}$. In this case, any of the definiens meets the substitutability condition with a qualified degree of exactness for the definiendum. The qualification of the degree of exactness is called *fuzzy conditionality*. The definienses are not exactly substitutes among each other. They may, however, be substituted among each other with a fuzzy conditionality.

Notice that the fuzzy definitional set, just like the crisp definitional set, is convex, closed and bounded. It is also a set-to-point mapping where we associate a degree of accuracy of meaning of each definiens to the definiendum. The condition of the crisp definitional set where $\mathfrak{F}(\cdot)$ is a set-to-point mapping with $\mu_{\mathfrak{M}}(\mathfrak{m})=1$, $\forall \mathfrak{m} \in \mathfrak{M}$ is linguistically unattainable and hence can be restructured to include a condition that there exists at most, one $\mathfrak{m}^* \in \mathfrak{M}$ such that $\mu_{\mathfrak{M}}(\mathfrak{m}^*)=1$. In other words, we transform the set-to-point mapping to a set-to-set mapping. If $\mu_{\mathfrak{M}}(\mathfrak{F}(\mathfrak{m})) \in (0,1]$ $\forall \mathfrak{m} \in \mathfrak{M}$ for the term $\mathbf{T}$ then we say that $\mathbf{T}$ is definable and if $\exists \mathfrak{m}^* \in \mathfrak{M} \ni \mu_{\mathfrak{M}}(\mathfrak{F}(\mathfrak{m}^*))=1$, then we say that $\mathfrak{m}^* \in \mathfrak{M}$ is a perfect explicator for $\mathbf{T}$. On the other hand, if $\exists \mathfrak{m}^{**} \in \mathfrak{M} \ni \mu_{\mathfrak{M}}(\mathfrak{F}(\mathfrak{m}^{**}))= \bigvee_{\mathfrak{m} \in \mathfrak{M}} \mu_{\mathfrak{M}}(\mathfrak{F}(\mathfrak{m}^{**}))$, then we say that $\mathfrak{m}^{**} \in \mathfrak{M}$ with an associated membership value $\mu_{\mathfrak{M}}(\mathfrak{F}(\mathfrak{m}^{**})) \in [0,1)$ is an inexact or fuzzy explicator for $\mathbf{T}$. The existence of fuzzy explicator extends the domain of the applicability of terms and concepts of the classical exact explicator.

**Lamma 6.4.2.2: A Fixed Point Lemma for Exact Symbolism**

An element $\mathfrak{m}^* \in \mathfrak{M}$ is a fixed point of a fuzzy definitional mapping in a language $\mathbf{L}$ for a given term $\mathbf{T}$ if and only if $\mathfrak{m}^*$ is a fixed point of the set-valued mapping $\mathfrak{F}: \mathfrak{M} \to \mathfrak{M}$ with $\mu_{\mathfrak{M}}(\mathfrak{m}^*)=1$ and $\mathfrak{m}^* \equiv \mathbf{T} \in \mathfrak{F}(\mathfrak{m}^*)$.

**Theorem: 6.4.2.1: A Fixed Point Theorem for Exact Symbolism**

If $\mathfrak{M} \subset \mho$ is a non-empty, compact and convex fuzzy definitional set for a term $\mathbf{T} \in \mho$ in a language $\mathbf{L}$, then the definitional function $\mathfrak{F}(\cdot)$ from $\mathfrak{M} \subset \mho$ to $\mathfrak{M} \subset \mho$ has a fixed point $\mathfrak{m}^* \in \mathfrak{M}$ such that $\mathfrak{F}(\mathfrak{m}^*)=\mathbf{T}$ with $\mu_{\mathfrak{M}}(\mathfrak{m}^*)=1=\mu_{\mathfrak{M}}(\mathbf{T})$, if and only if $\mathfrak{m}^* \equiv \mathbf{T} \in \mathfrak{M}$ and the definiendum, $\mathbf{T}$ is identical to the definiens $\mathfrak{m}^*$ as an exact definiens.

The theorem of definitional exactness in relation to a fixed point in the definitional set suggests that the condition of exactness of symbolic representation is that the definiendum (symbol) is an element of the definitional set containing all the definiens in which case $x$ is $x$ and if $y$ is a definitional symbol of $x$, then $y \equiv x$.

In this case, the fixed point of the definitional function is the perfect explicator where in this case $\mathbf{T}$ is a linguistic primitive. In general, the explicator set is contained in the definitional set. To examine the requirements of definitions in scientific practices in any given language and any area of knowledge production, we specify the following set of axioms call the *axiom of definability* and *explicability*.

### 6.4.3 Definability and Explicability Axioms in the Language L

**Axiom 1: Interchangeability**

If $\mathfrak{m}$ is a definiens for the definiendum $\mathbf{T}$, then the definiens $\mathfrak{m}^* \equiv \mathbf{T} \in \mathfrak{M}$ must serve as a good substitute for the definiendum $\mathbf{T}$ in the language $\mathbf{L}$ such that if $\mu_{\mathfrak{M}}(\mathfrak{m}) = \alpha$ is a measure of the degree of definitional exactness, then $\alpha \in (0,1]$ and the definiens $\mathfrak{m}$ can be substituted for the definiendum $\mathbf{T}$ with an $\alpha \in (0,1]$ degree of exactness in the language $\mathbf{L}$.

**Axiom 2: Non-tautology or non-circularity**

The definiendum must not occur as a definiens or the definitional set must not contain the definiendum. Similarly, the explicandum must not occur as an explicatum or the explicator set must not contain the exlicandum

**Axiom 3: Non-ambiguity**

There must be one definiendum (explicandum) that corresponds to the definitional set (explicator set). In other words, many definienses (explicata) can correspond to the definiendum (explicandum), but there is one and only one definiendum (explicandum) where the definiens (explicatum) can serve as a good or exact representation of the definiendum (explicandum).

**Axiom 4: Non-contradiction**

The equality between the definiendum and the definiens or between the explicandum and the explicatum must be such that there should be no contradiction in separate definitional or explicator sets. Thus, if there are two or more definitional (explicator) sets for a definiendum (explicandum), then the definitional (explicator) sets are equal. In other words, the same definiens (explicatum) must not serve as a representation of different definienda (explicanda), and if it does, then the definienda (explicanda) are the same.

**Axiom 5: Functional continuity**

The definitional function must be of a continuous mapping over the definitional (explicator) set through its membership characteristic function so as to rank the definienses (explicata) in degrees of exactness of meanings associated with the definiendum (explicandum), where each degree of exactness is bounded in the [0,1] interval. This is the meaning of continuum in languages.

Axioms 1-5, the fixed point lemma and the theorem on the fixed point-mapping of exactness provide us with the conditions under which exact symbolism in the information-knowledge structure can be created to support exact logic and exact mathematical reasoning in the mathematical language, $\mathbf{L}$. These axioms serve as the conditions for constructing a definitional mapping, a definitional set, explicator function and explicator set in any given language and knowledge sector for both fuzzy and exact symbolic systems. It may also be reflected on the notion that the primary principles in establishing definitional and explicative structures of empirical and axiomatic theories in knowledge systems, including relational structures of derived and corresponding conceptual sub-systems, in order to provide logical closures and the rules in introducing new epistemic elements into the knowledge system, form an important foundation of the epistemic process with its grammars of thought. From these axioms, we state an impossibility theorem for exact symbolism in the information-knowledge representation.

### Theorem 6.4.3.1: The Non-existence of Exact Symbolism

Given the axioms 1-5 and the conditions of definitional mapping of $\mathcal{F}:\mathcal{M}\to\mathcal{M}$, there does not exist a definitional set $\mathcal{M}$ that contains $m^*$ such that $\mathcal{F}(m^*)=\mathbf{T}$ with $\mu_\mathcal{M}(m^*)=1$, where $\mathcal{F}(m^*)=\mathbf{T}\equiv m^*$ in any language, $\mathbf{L}$, to provide conditions of exact symbolism in any knowledge area. Hence an exact symbolism of the information-knowledge representation with a defined content is impossible and only fuzzy or vague symbolism is possible.

### Proof

Consider the definitional set $\mathcal{M}$ and by the construct, $\mu_\mathcal{M}(\mathcal{F}(m))\in[0,1]\ \forall m\in\mathcal{M}\subseteq\mho$. For all $m\in\mho$, if $\mu_\mathcal{M}(\mathcal{F}(m))=0$ then $m\in\mho\Delta\mathcal{M}$ (where $\Delta$ is the relative complement of $\mho$ and $\mathcal{M}$) but $m\notin\mathcal{M}$. Similarly, if $\mu_\mathcal{M}(\mathcal{F}(m))\in(0,1]$ then $m\in\mathcal{M}$ and $m\notin\mathcal{M}\Delta\Omega$. By the axiom of non-tautology $\mathbf{T}\notin\mathcal{M}$ and the definitional set is closed and bounded in the unit simplex where $\mathcal{F}(m)\in(0,1)$ and hence there does not exist $m^*\in\mathcal{M}\ni\mu_\mathcal{M}(\mathcal{F}(m^*))=1$ with $m^*\equiv\mathbf{T}$. If there exist $m^*\in\mathcal{M}\ni\mu_\mathcal{M}(\mathcal{F}(m^*))=1$ with $m^*\equiv\mathbf{T}$, then $\mathbf{T}\in\mathcal{M}$ creating circularity that violates the axiom of non-tautology. Since there is no set $\mathcal{M}$ that contains $m^*\in\mathcal{M}\ni\mathcal{F}:\mathcal{M}\to\mathcal{M}$ with $\mathcal{F}(m^*)\equiv\mathbf{T}\equiv m^*$ and $\mu_\mathcal{M}(\mathcal{F}(m^*))=1$, we conclude that exact symbolism in information-knowledge representation is not possible and only vague or fuzzy symbolism is possible with $\mu_\mathcal{M}(\mathcal{F}(m))\in(0,1]$                                    □

### Note 6.4.3.1

The corollary to the impossibility theorem for exact symbolism in an information-knowledge construct is that the only inputs available to cognitive agents, operating over the epistemological space, are defective information structures that constrain exact symbolism and exact logic. In other words, the epistemic existence of perfect information structures, as inputs into the decision-choice processes in the knowledge-production system, is impossible. This corollary to the *impossibility theorem* of the exact symbolism, viewed in terms of information, is consistent with analytical work and conclusions by Russell, as a formalist, that we have previously quoted, that of the intuitionists, such as Brouwer that we have also quoted, as well as those conclusions of Max Black [R8.55], the discussions of the philosophical basis of intuitionist logic by Dummett [R14.3], discussions by Max Planck on exact science [R8.55] and the essays in Benacerraf and Putnam [R14.5].

The discussions of exact and inexact symbolism are extremely important to our knowledge-development enterprise as viewed from formal and informal information representation that must be operated on with either exact or inexact logic. Any position taken will also reflect on digital-analog and discrete-continuum debates. All these discussions extend to the concepts of dualism and duality in relations to the paradigms of thought and how the opposites of polarity and duality are relationally examined and connected to create laws of thoughts. They also relate to how contradictions are not accepted as truth value with the principle of excluded middle and how contradictions are accepted as truth value on the principle of continuum. The laws of thought that emerge define conditions to accept conclusions on the basis of true-false relations. The nature of acceptance and rejection of either truth or falsity depends on exactness and inexactness of information representation and reasoning to create thought.

The methodological point that must be brought into focus is that the concepts of digital, discrete, analog and continuum are linguistic tools that in turn point to the nature of analytical tools used in the epistemological space in other to claim knowledge or validate the truth of propositions that are essential to the organization and management of the social collectivity. Viewed in a set-theoretic notion and categories of being, every digit has an analog covering that defines its existence and meaning just as every discreteness has a continuum covering while every exactness has a fuzzy covering to make sense of the language of knowledge. Every digit or discrete is a point set or a singleton set. Every analog or continuum is a set or an interval. The analytical results of the digital and discreteness, even though are assigned exactness by predetermined logical rules outside the cognitive agents, are contained in the sets of analog and continuum which carry with them inexactness and whose clarities are decision-choice determined as part of the knowledge-production process to deal with quality and complexity. The digital and discrete methodology avoids the controversies of subjectivity, complexities of qualitative dispositions and concentrates on quantity and time. The analog and continuum methodology embraces controversies of subjective judgment and complexities of quality in dealing with the knowledge production. The analytical result is that the division of science between exactness and inexactness cannot be sustained by exactness of logical claims over the epistemological space. The universal principle is that, the defective information structure, carrying qualitative properties of fuzziness is a characteristic of all knowledge areas and sciences.

# Chapter 7
# Ontology, Epistemology, Explication and Exactness in Mathematics and Sciences

All the questions raised in the previous chapters fall under either the set of questions of ontological type or the set of questions of epistemological type. Most of the questions are of epistemological type. The theory of the knowledge square is a meta-theory on epistemology and knowledge production. The point of entry of the meta-theory is that the information input for knowledge production is defective due to fuzziness and incompleteness that create qualitative and quantitative inexactness for all sciences. Fuzziness is associated with qualitative disposition of information while incompleteness is associated with quantitative disposition of information about epistemic objects with neutrality of time. The point of departure is a search for an appropriate symbolism that will incorporate fuzziness and incompleteness of information with a further search for appropriate logic of operations that will allow exact equivalences to be abstracted from inexactness. All the debates in the last analysis are about what we claim to know about *what there is* in the ontological space and the process of knowing. In the epistemic process of knowing, the solution to the problem of exactness forces us to choose between two principles that must be taken as intuitive assumptions. They are the ontological principle of exactness and inexactness, on one hand, and the epistemological principle of exactness and inexactness on the other. The inter-relational structures between ontology and epistemology regarding exactness and inexactness are displayed in Figure 7.0.1. In the process of the knowledge production, are we willing to assume the ontological principle of exactness about the elements in the universal object set as defined in the potential space relative to knowing? In other words, are we willing to accept the fundamental principle that ontological objects, processes and states are exact, and hence there are no vague objects and there is no defective ontological information structure associated with them as in Cohort II? Alternatively, are we to accept ontological vagueness and epistemic exactness as in Cohort III? Another choice that is to be faced is that of accepting complete exactness in both spaces where the epistemological information is the exact replica of the ontological information as in Cohort I. The final choice is to accept complete inexactness in both the spaces with vague ontological and epistemic objects as in Cohort IV. Let us examine these in a little more detail.

K.K. Dompere: The Theory of the Knowledge Square, STUDFUZZ 289, pp. 117–138.
springerlink.com                    © Springer-Verlag Berlin Heidelberg 2013

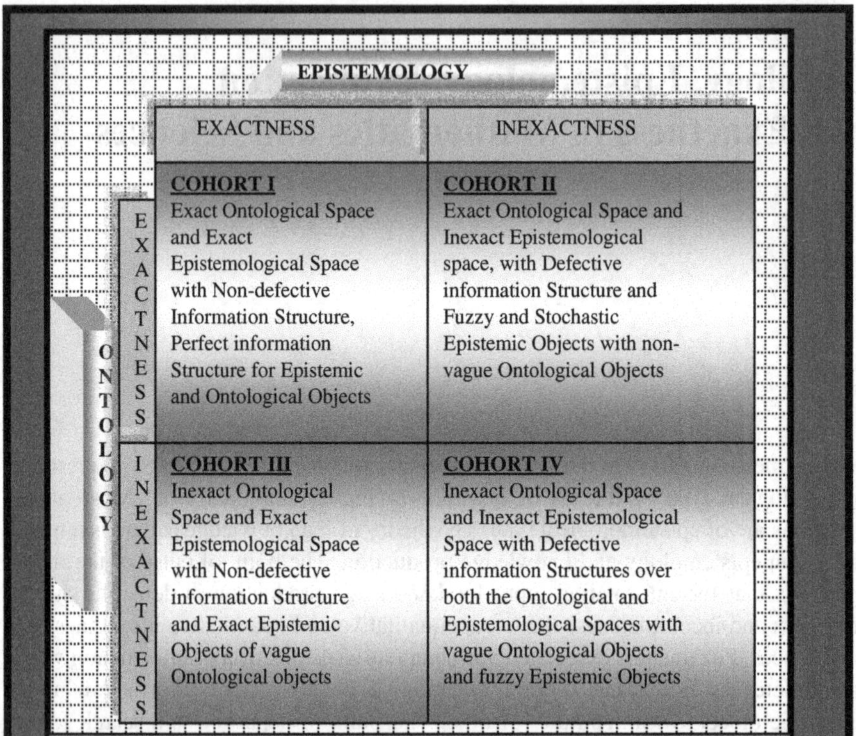

| EPISTEMOLOGY | | |
|---|---|---|
| | EXACTNESS | INEXACTNESS |
| E X A C T N E S S | **COHORT I**<br>Exact Ontological Space and Exact Epistemological Space with Non-defective Information Structure, Perfect information Structure for Epistemic and Ontological Objects | **COHORT II**<br>Exact Ontological Space and Inexact Epistemological space, with Defective information Structure and Fuzzy and Stochastic Epistemic Objects with non-vague Ontological Objects |
| I N E X A C T N E S S | **COHORT III**<br>Inexact Ontological Space and Exact Epistemological Space with Non-defective information Structure and Exact Epistemic Objects of vague Ontological objects | **COHORT IV**<br>Inexact Ontological Space and Inexact Epistemological Space with Defective information Structures over both the Ontological and Epistemological Spaces with vague Ontological Objects and fuzzy Epistemic Objects |

(Left vertical label: ONTOLOGY)

**Fig. 7.0.1** Descriptive Relations of ontological and Epistemological Information Structures and the Corresponding Objects of Knowing

## 7.1   Ontology, Epistemology and the Defective Information Structure

The defective information structure (also imperfect information structure) is characterized by vagueness, ambiguities and incompleteness. Under this principle of ontological exactness, objects, processes and states are what they are, and hence the questions of exactness and inexactness ontologically disappear, in that the ontological space is characterized by perfect information structure which cognitive agents seek to know from the activities in the epistemological space. We must then move on to the epistemological principle of either inexactness or exactness relative to the ontological principle of exactness. This limits us to Cohorts I and II in the knowledge-production process. In this case, the questions of inexactness and exactness characterize the epistemological space as well as involve the logical problems of cognition and the methodology of knowledge production. We have exact ontological objects that exist outside the creational zone of cognitive agents. Similarly, we have the exact and inexact epistemic objects that are created by cognitive agents through an information sending-receiving module combined with decision-choice actions over the epistemological space.

The problem confronted by cognitive agents is to answer the question as to the degree of isomorphism that exists between each ontological element and the corresponding epistemic elements. This is the epistemic problem of knowing and knowledge discovery. All the epistemological problems must be settled with decision-choice actions in the epistemological space by cognitive agents with an input-output process. The logical and methodological problems involving vagueness and incompleteness are seen in terms of defectiveness at the level of the information structure while the ambiguities and approximations are seen in terms of defectiveness at the level of thought. All of these are seen in terms of the philosophical basis of information-knowledge representations and reasoning through the nature of an accepted symbolism and manipulations at the presence of defective information structure in the epistemological space. The epistemological space presents us with opportunity to create the technological facility that can take information as an input, process it into knowledge as an output, and then check it for the degree of closeness between ontological and epistemic objects. The distance between an ontological object and corresponding epistemic object is defined as the *knowledge distance* (or an *epistemic distance*) that can be asymptotically closed by increasing cognitive action. This input-output transformation technology from information to knowledge has always be a source of controversy as well as a power of human existence.

Over the epistemological space, there are a number of situations that knowledge finders find it convenient to assume the existence of Cohort I of Figure 7.1.1. This is taken to be an ideal case with perfect epistemological information. This case has a number of problems in terms of how one can justify the perfect epistemological information structure except by appealing to the principle of analytical convenience through the use of assumption. Most knowledge finders have their search activities taking place in Cohort II. Let us take a look at the information structure of Cohort II which is presented as in Figure 7.1.1. In this diagram, attention is on the possible combinations of qualitative and quantitative dispositions of information to create the information structure for symbolic representations and epistemic processing. Four combinations relating to vagueness and completeness of information are presented for the creation of the information structure as an input for epistemic processing. They are a) qualitative defectiveness and quantitative non-defectiveness with non-stochastic fuzzy epistemic objects (Cohort I of Figure 7.1.1), b) qualitative non-defectiveness and quantitative non-defectiveness with non-fuzzy non-stochastic epistemic objects (Cohort II of Figure 7.1.1), c)qualitative defectiveness and quantitative defectiveness with fuzzy stochastic epistemic objects (Cohort III of Figure 7.1.1), d) qualitative non-defectiveness and quantitative defectiveness producing stochastic non-fuzzy epistemic objects (Cohort IV of Figure 7.1.1).

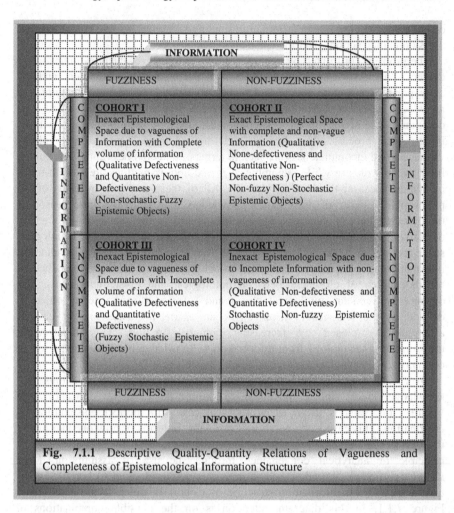

**Fig. 7.1.1** Descriptive Quality-Quantity Relations of Vagueness and Completeness of Epistemological Information Structure

In this distribution of possible information structures, the dominant practice is to assume away the conditions of fuzziness and claim exactness of incomplete information structure in order to create non-fuzzy defective information structure established by quantitative incompleteness, where the epistemic objects are stripped of their qualitative characteristics in order to work in the exact epistemological space. The analytical difficulties in this case have been explained through the nature of ordinary language and the principles of acquaintance to create experiential information structure which is sometimes referred to as the sense data. This is Cohort IV in Figure 7.1.1. There is a need of increasing importance to deal with the simultaneity of fuzziness and incompleteness of the defective information structure in the Cohort II of Figure 7.1.1. This need is taken as an essential reference point to examine other analytical frameworks that may be carved out from this Cohort.

Alternatively, if we impose the principle of ontological inexactness as our initial condition of cognition on the system of the knowledge search, and knowledge production, then the elements in the universal object set come to us in vagueness in the epistemological space. In this situation, we are restricted to the Cohorts III and IV with vague ontological elements. In this case with vague ontological elements, the epistemological problem of exactness is simply *a phantom one* that is encapsulated in ontological inexactness. Since objects, processes and states in the ontological space are taken to be vague, no analytical and logical process can be constructed by cognitive agents to achieve an epistemic exactness, and hence the claim to exact science is nothing but a fiction of cognitive delusion, and the justification for the use of the classical paradigm for knowledge search and production falls apart. Those thinkers who claim the ontological existence of vague objects, processes and states must offer some justified principle for the claim of exact science. They must also show how the vague ontological objects are transformed into exact epistemic objects in the epistemological space.

The practice of exact science in the epistemological space is not possible given a space of ontological inexactness where ontological objects reside under conditions of vagueness. Elements in the universal object set as abstracted from the space of cognitive potential (the ontological elements) are what they are, and the characterization of them as vague by human cognition seems to be logically wanting. Exact science and the existence of ontological vagueness (ontological imperfect information structure) are incompatible with the epistemological space at the level of a search for exact epistemic objects in sciences. Digital and discrete information representations are meaningless if the ontological objects are vague, and hence methodological discreteness and digital processes are of no help to epistemic activities in the epistemological space.

When we accept the ontological principle of non-vague objects, processes and states of elements in the universal object set (that is, when we accept the basic principle of ontological perfect information structure), then the problems of concepts of vagueness, inexactness and incompleteness in information, discrete and continuum in methodology and paradigms of thought are shifted from ontology to epistemology, in that the conceptual problems do not depend on the natural order of things but rather they depend on the epistemic order of things. The problem of defective information structure is defined and solved in the epistemological space.

The epistemic order of things depends on human perception and decision-choice actions in access to ontological information signals by acquaintances, classifications and category formations, symbols of representation, ability to know, methods of cognition and criteria of knowledge verification and acceptance. Here, exactness, inexactness, vagueness and ambiguity become qualitative properties of the epistemic items and the knowledge structure but not properties of ontological items. In this way, the epistemological space is characterized by defective information structure that we may also term epistemologically imperfect information structure with phantom candidates for knowledge search that reside in Cohort II of Figure 7.1.1. At the level of epistemology, the concepts of exactness and inexactness exist in duality under cognitive tension that generates the energy for change from epistemic inexactness to epistemic exactness in a continuum. Such a cognitive tension is resolved by logical transformation between extremes of opposites.

The logical transformations work through categorial conversions under the creative forces of categorial moments that provide the forces of conversions from varying stages of epistemic inexactness to varying stages of epistemic exactness through the decision-choice rationalities of cognitive agents. For the conversion process to work, we must identify and establish two categories of a set of characteristics that defines an epistemic *category of exactness* and another set of characteristics that also defines an epistemic *category of inexactness.* We must keep in mind that cognition involves the ontological trinity of matter, energy and information which are transmitted from the ontological space and received by cognitive agents as defective information structures in the epistemological space from which exactness is sought from inexactness through a process. All models of knowledge production matter, energy and information are the basic elements that are defined in categories of polarity, duality and continuum. They are connected in a transformational continuum where matter is energy; energy is information; and information is matter in the system of categorial conversions as shown in Figure 7.1.2. The concept of exactness must be seen as a fuzzy characteristic set with a membership characteristic function. Corresponding to it is its complement of the concept of inexactness with a membership characteristic function. Any epistemic object is defined by the union of the two sets to create the identity of the epistemic object which is under transformational activities of the set of inexact-exact combinations of the defining characteristic sets altering its quality and hence, its identity.

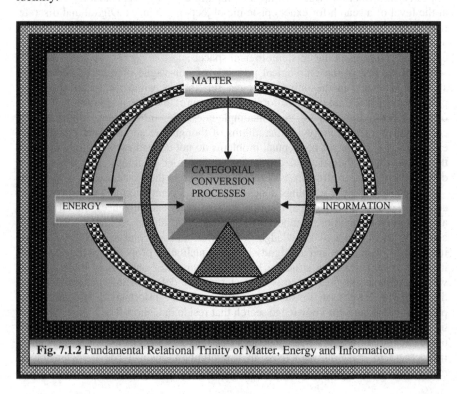

**Fig. 7.1.2** Fundamental Relational Trinity of Matter, Energy and Information

### 7.1.1 Inexactness, Exactness and Categorial Conversion in the Epistemological Space

The process of categorial conversion is to introduce a categorial moment that acts on the inexact category and transforms it into an exact category by changing the relationship between exactness and inexactness as they relate to epistemic items. Here, we are concerned with the process of *epistemic categorial conversions* and corresponding *categorial moments* as opposed to the process of *ontological categorial conversions* and *ontological moments*. At this point it is useful to keep in mind that there are two transformations taking place through the forces of categorial conversions with corresponding categorial moments and two sets of conditions of categorial convertibility. We have ontological *categorial conversion* with ontological *categorial moment* and the set of conditions for *categorial convertibility* all of which are outside the control of cognitive agents. We also have the epistemic categorial conversion with *categorial moment* and a set of conditions of *categorial convertibility*, all of which are under the control of cognitive agents. The epistemic categorial conversion is a vehicle to find out and understand the nature and structure of the ontological process. Given the events in the ontological space, the task in the epistemic conversion is to create a cognitive *input-output technology* in the epistemological space that will be used in transforming the defective information structure into dependable knowledge. The cognitive *input-output technology* is the paradigm of thought with its supporting components.

In the epistemological space, any epistemic item presents itself in an exact-inexact duality. Any epistemic item or process or state is said to be exact when the set of characteristics of exactness in the category is judged to be greater than the set of characteristics that defines the inexact category by some rule of human decision-choice action. At the level of categorial formation, the exactness and inexactness are seen as a cognitive duality under logical tension. At the level of categorial conversion, the inexactness and exactness are seen as properties of ignorance that are due to lack of some knowledge that is attributable to *defective information structures* relative to the elements in the ontological space. The elements in the universal object set are sharp and exact in the sense that *what there is*, is *what there is* as the ontological elements.

The process of knowing involves dealing with inexactness due to epistemological defective information structure where we assume the existence of exact ontological elements and begin the path of knowing with inexact epistemic objects that are continually being refined towards exactness as new information is revealed. The process of creating exact epistemic elements from inexact epistemic elements operates on the basis of human decision-choice actions in the process of knowing from the information-perfect ontological space to the information-imperfect epistemological space that provides us with a separation principle between the categories of exactness and inexactness. The cognitive process is such that the claim of exactness must satisfy the *fuzzy conditionality* and the *stochastic conditionality* in the categorial conversion process as we work in the Cohort II in Figure 7.1.1. The fuzzy conditionality is a possibilistic qualification in dealing with the qualitative disposition to subjectively interpret vagueness and degree of acceptability. The stochastic conditionality is a probabilistic qualification in dealing with quantitative disposition to subjectively interpret approximations and degree of acceptability. Both of those are expressed on the epistemic belief of exactness in processing the defective information structure over the epistemological space.

On this basis, the reflection here is that, from the onset we must dismiss the principle of ontological claim that vagueness and inexactness may be properties of the elements in the universal object set (the potential space for knowing). All ontological objects, processes and states are what they are and exist under exact conditions of their natural setting. All discussions must, therefore, center on inexactness, exactness, vagueness and ambiguity as properties of epistemic objects, processes and states, and the role they play in the cognition of ontological elements of the universal object set. It is these properties of epistemic elements that create conditions of ignorance-knowledge duality and the dynamics of knowing. The structures of exactness and inexactness of epistemic elements reside in defective information structure, processing of epistemologically defective information structure and faulty reasoning or inefficient laws of thought in the geometry of knowing. In this discussion, we arrive at a conclusion that exactness is a property of the potential space (universal object set) and inexactness is a property of the possibility space, the probability space and the space of the epistemic actual in the process of knowing the elements in the potential space. Any claim of exact knowledge in the epistemological space is simply conditional on decision-choice actions on the part of the cognitive agents as well as conditional on the proportion of inexactness contained in it as seen from the structure of *fuzzy logical conditionality, stochastic logical conditionality* and *fuzzy-stochastic logical conditionality*. The existence of the fuzzy logical conditionality leads to the development of possibilistic (fuzzy) statistics. The existence of stochastic logical conditionality leads to the development of both classical and Bayesian statistics. The existence of fuzzy-stochastic logical conditionality leads to the development of generalized fuzzy-stochastic statistics.

The assignment of qualities of vagueness, ambiguity, inexactness and exactness as properties of epistemic objects provides us with a useful entry point into the discussions on the nature of *exact science* and *inexact science* under the principles of duality and continuum that provide us with a separation tool through decision-choice actions in the epistemological space and the knowledge-production enterprise. Exactness and inexactness are not characteristics of natural processes; they are characteristics of the human cognitive process in the epistemological space. The entry point is only analytically useful when we view science as part of the global knowledge-production process and the scientific enterprise is viewed as an aspect of the knowledge-production enterprise both of which involve human actions in social production. The knowledge-production enterprise is an organization in support of the process of knowing as well as the process of economic production.

A question, thus, arises as to whether this act of knowing is a mental process or a physical process or both. To what extent can exactness or inexactness be claimed to characterize a knowing process within the general cognition? It may be emphasized that the postulate that inexactness or vagueness is a property of ontological elements in the universal object set renders the existence of exact and inexact sciences meaningless. Under this postulate, epistemological exactness cannot be found to clean up ontological inexactness no matter how hard we try to refine thinking and the composite toolbox of cognition. The problem of epistemic exactness under conditions of ontological inexactness becomes a *phantom one* as we have previously stated. Let us keep in mind from the theory of the knowledge square discussed in the previous chapters, that the knowledge-production process is a cognitive mapping from the epistemological space to itself by the methodological constructionism, then from the epistemological space to the ontological space by the methodological reductionism, and further from the ontological space to the epistemological space by constructionism.

The postulate of exactness of ontological elements ensures a meaningful categorial formation, development of meaning, creation of definitions and production of relations of things in the universal object set by any common language that can be explicated for the knowledge production in the epistemological space. This may be called a linguistic category. Such a process is not possible under the postulate of inexactness or vagueness of ontological elements in the universal object set. In the language formation and knowledge production, we have ontological elements that exist in the state of exactness under natural substitution-transformation processes outside the awareness of cognitive agents. They exist as what they are in their environment of their being. As we have argued, the number of *ontological categories* is infinite, and infinitely closed under internal categorial conversions in nature itself. Given the exact ontological categories, there are also the *epistemic categories* that exist in the state of inexactness and open under cognitive transformation of logical categories. The epistemological categories are related to the ontological categories through the creation of *linguistic categories*. The discovery of epistemic items and the cleaning up of the vagueness and ambiguity that allows equality to be established between ontological elements and epistemic elements are the works of the knowledge-production process and its enterprise. This knowledge production process works with the cognitive processing machine of defective epistemological information structure to derive conditional epistemic items.

The number of epistemic categories are always finite, the size of which varies in time and expands with an increasing knowledge base as we resolve conflicts in exact-inexact duality for different epistemic elements. One may view the knowledge-production process as a set-to-point mapping. To see this, consider the ontological space as $\mathfrak{U}$ with generic element $\pi \in \mathfrak{U}$. The element $\pi \in \mathfrak{U}$ is called an ontological item. Now define an epistemological space, $\Psi = \mathfrak{U}$, with a generic term, $\eta \in \Psi$, and the space of the process of knowledge search $\varphi \in \Phi$. The element $\eta \in \Psi$ is called an epistemic item while the element $\varphi \in \Phi$ is called cognitive transformation function. For each ontological element, $\pi \in \mathfrak{U}$, there corresponds a category of epistemic items, $\Psi_\ell$ which may be viewed as belonging to the possible world set, $\Psi_1 \subset \mathfrak{P} \subset \mathbb{P}$ through $\varphi \in \Phi$. First, we may note that given $\mathfrak{U}$ we have

$\Phi : \Psi \to \Psi$ and if $\pi \in \mathfrak{U}$, then there exist $\Phi_i : \Psi \to \Psi_i$ with $\mu_{\Psi_i}(\eta) = \sigma_i \in [0,1]$, and where $\Psi_i = \left\{ \left( \varphi_i(\eta), \sigma_i \right) \mid i \in \mathbb{N} \text{ and } \varphi_i(\eta) = \pi \in \mathfrak{U}, \varphi_i \in \Phi_i \right\}$ with a set of claimed knowledge about $\pi \in \mathfrak{U}$ with an aggregate degree of confidence of $\mu_{\Psi_i}(\eta) = \sigma_i \in [0,1]$ where $\mathbb{N}$ is an index set of cognitive transformation functions such that $i \in \mathbb{N}$. The construct of the aggregate degree of confidence is the aggregate epistemic conditionality which is obtained as a sum of fuzzy conditionality and stochastic conditionality. Let us note that, $\Phi$ is a set of cognitive processing machines, with imperfect information input, to create categorial moments for categorial transformations in the epistemological space.

In order to speak of exactness and inexactness of science, we must first have a notion of what science means and what are the defining characteristics of an element in one of our linguistic categories that allow an entry of such an element into an epistemic category. What characteristics distinguish science from non-science? In other words, what are the characteristics that demarcate science from non-science? Science, itself, belongs to a linguistic category that is used to examine the relationship between epistemological categories and ontological categories as well as to create associations between ontological elements and epistemic elements. Science is neither ontological nor epistemological. Science is simply a linguistic variable with degrees of linguistic values in meaning that help the cognitive agents to connect the elements in the epistemological space to some elements in the ontological space. As a member of linguistic category, it qualifies to be characterized as exact from inexact after explication which is decision-choice determined. When the linguistic concept of science has been explicated, then we can deal with any qualification such as exact, political, medical, and others that may be attached. Here, the whole concept of science is under microscopic investigation as to how science differs from non-science in our linguistic categories. At the level of knowing, how does science differ from other approaches to knowledge production? The point here is that we cannot speak of a scientific theory, methods of science, scientific discovery, exact science, inexact science and other exotic terms and phrases that involve science when we do not even have a collective agreement on what constitutes science and its linguistic category. The collective agreement of the conceptual content of science requires us to find a solution to the explication problem of science.

The solution to the problem of explication of science and its linguistic category is very important in solving the methodological problems of the knowledge-production process and scientific discovery. It will allow us to form a reasonable category of knowledge areas and answer questions of general interest. Are psychology, economics, mathematics, symbolic logic, engineering and medicine members of categories of science? What areas of human knowledge or quest for knowing are not sciences? Our interest here is not on the parameters of knowledge that we may call science but rather on the problem of exactness and inexactness of science. The problem can only be meaningfully dealt with when the concept of science has been explicated. Viewed from explicator and definitional sets from Chapter 6, the explication problem relates to the decision-choice problems between the classical and the fuzzy paradigms of thought in dealing with the information-knowledge production process. The relationship, whether in difference or similarity in thinking to support one paradigm or the other, consists in assigning to the world of knowledge the linguistic properties that we hold to form categories for information processing.

## 7.2 Science and Non-science as Linguistic Categories in the Information-Knowledge Production Process

In order to know what elements to place in each category of science, we must provide a working definition of science that will help in establishing the parameters of decision-choice action on the knowledge areas. We shall discuss a few interrelated explications that will provide some content to the linguistic category of science and hence that of non-science. Some views define science as logical processes of formulating models in *natural sciences*. These models must predict outcomes under conditions of *uncertainty* where the uncertainty is of stochastic type (for discussions on types of uncertainty see [R6] [R6.29] [R11.21]). The predicted outcome is then tested against the available data to see if the future *predictions* agree with the past and thus connected to the present to establish a continuum process. We shall refer to the past-present-future connection as the *time trinity* (this time trinity is best described by *sankofa bird* in the *Akan* philosophical system in Ghana, which simply states that by retrieving past information and connecting it to the current information one can assess the future information regarding directions of events [R14.29].).

### 7.2.1 Characteristic Sets for Definition and Explication of Science

In the explication process respect, there is an underlying system from which the potential space is being approximated by models to connect to the epistemic actual by cognition through decision-choice processes. The potential space exists independently of human understanding, and hence reflects objective reality with perfect ontological information structure in that it is what is there. The existence of the known actual depends on human conceptualization and thought processes, and hence reflects the subjective reality or an epistemic reality. There are a number of approaches to establish the defining characteristics of the subjective reality in terms of science as a linguistic category over the epistemic space through cognition. One such approach produces the following list as the characteristics of science:

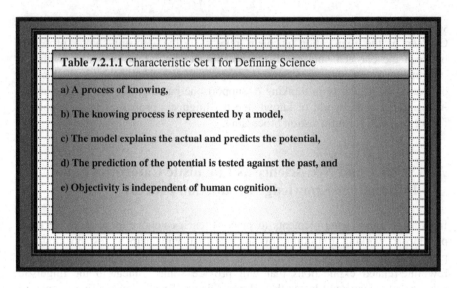

**Table 7.2.1.1** Characteristic Set I for Defining Science

a) A process of knowing,

b) The knowing process is represented by a model,

c) The model explains the actual and predicts the potential,

d) The prediction of the potential is tested against the past, and

e) Objectivity is independent of human cognition.

Here, a question arises as to the meaning of a model. The concept of a model must in turn be explicated if the linguistic category of science is to be useful for analytical work. The concept of natural science imposes restrictions on the areas of the modeling. It may be kept in mind that while subjectivity is independent of *what there is*, the knowability of *what there is* a subjective phenomenon and hence dependent on the knower. It is this dependency and the relationship between knowability and subjectivity that give content and meaning to the question: *what does the knower know and how does the knower know that he or she knows*? Alternatively stated, the existence of ontological elements is independent of the existence of cognitive agents, while the existence of epistemic elements is firmly dependent on the existence of cognitive agents.

Alternatively, science has been seen as the effort to discover and increase human understanding of how physical reality works. This effort is exercised through the provision of an alternative set of characteristics on the basis of which science may be defined and explicated. These characteristics are provided in Table 7.2.1.2.

From such basic characteristics of science as a linguistic category, a question arises as to whether science is defined by its methods of obtaining knowledge or exactness in its content of knowledge or defined in terms of the knowledge area. To answer this question, methods associated with science are reviewed as a framework that allows claims to knowledge to be verified as science through the characteristics set as presented in table 7.2.1.3.

**Table 7.2.1.2** Characteristic Set II for Defining Science

a) Controlled methods for data collection by observing physical evidence of natural phenomena;

b) Construction of theoretical explanation on the basis of data analysis as to how natural phenomena work in the sense that they describe human understanding of the physical processes for prediction;

c) The prediction is based on explanatory rationality, and experimental and empirical research under the methods of generalized hypothesis and tests of these hypotheses under controlled conditions.

**Table 7.2.1.3** Characteristic Set III for Defining Science

a) Reproducible experiment or generally accepted technique

   that minimizes subjectivity under controlled conditions;

b) Observation of the object of phenomenon of interest;

c) The development of a set of empirical evidence from

   experiment or the observation;

d) Presenting the phenomenon in terms of a logical or

   mathematical exact representation from which hypotheses

   are developed, and claims of knowledge are made;

e) Peer review of the mathematical model, the experimental

   results and the drawn conclusions.

## 7.2.2   Explanation, Definition and the Grammar of Knowledge Areas

The linguistic terms of theory and model must be clarified for us to proceed in the explication of linguistic category of science. We must then relate them to the presence of defective information structure that generates uncertainties, risks and general systemic risk in the knowledge-production system.

### Definition 7.2.2.1: Theory

A theory is a logical system of reasoning, with a given information structure, under defined rules of thought, that describes (instructs) through explanation (prescription), and predicts through analyses or syntheses of the behavior of a phenomenon from which hypotheses (decision-choice rules) are advanced for empirical verification or falsification or corroboration or all of them as a test of validity for acceptance (practice). This definition relates to both explanatory and prescriptive theories.

### Definition 7.2.2.2: Model

A model may be viewed as a frame that represents the essential structure of a theory, in terms of mathematics and logic that explicitly project the maintained hypotheses in the theory for further empirical or sensitivity analysis . It is, therefore, an essential skeleton of the theory in a self-contained symbolic or mathematical language.

A model, just like a theory, may represent either explanatory conditions of the epistemic actual as revealed by existence or the past  relative to the present, or prescriptive conditions of the epistemic potential as contained in the future relative to the present. Here, an important distinction between prediction and prescription is required. While prediction say something about the repetition of the past in the future relative to the present under controlled conditions of a theory, prescriptive model may, while saying something about the future,  have no immediate connection to the past relative to the present in terms of information structures. It is in this connection that the elements in the possible-world set acquire conceptual and analytical meaning in the epistemological space and find their associations with the possible, the probable and the actual. Both prediction and prescription, however, may not relate to the past. Prescriptive theory is a present-future phenomenon that may not relate to the past. Explanatory theory is a past-present-future phenomenon while prediction is a past-future phenomenon [R2.9] [R11.22].

In terms of cognition, a model is, thus, contained in a theory without which a model acquires no logical authority and analytical consistency. Even statistical models of correlation, data analysis and others have theories behind them even though such theories may not claim an explanatory or causal rationality. A theory is thus a support of a model even though we encounter terms like theoretical model and empirical model that sometimes create intense debate and confusion in the knowledge-production process and claims to knowing. Explanatory model is to explanatory theory as prescriptive model is to prescriptive theory. Mathematics is used to express the essential features of the theory as a frame. Logic or laws of thought is used to manipulate the essential theoretical links of the model that connects both the theory and the model in the time trinity of past, present and future in the case of explanation, or present and future relative to the past in the case of prescription that is not explanatory based. Prescriptive theory has a connection to the possible world set and explanatory theory has connection to the possibility set.

Given the mathematics and the logic rules, computability allows a computing power to simulate different environments, as established by information structures in the epistemological space and captured by the relevant parameters, for the mathematical-model representation of the theory. The information in the model is summarized by a set of parameters that defines the applicable area of the theory and hence the model. Such parameters may be assumed to be exact or inexact (fuzzy) in defining the framework of mathematical-model representation. When the parameters are assumes to be exact in the epistemological space, we claim exactness of knowledge with no epistemic conditionality or with stochastic conditionality. When the parameters are inexact in the epistemological space, we designate inexactness of knowledge with either fuzzy or fuzzy-stochastic conditionality.

In all these searches for knowledge in the epistemological space and a search for an optimal path to it, philosophy has a critical role to play to help in distinguishing knowledge from fiction and simple opinions as well as to distinguish science from non-science. Furthermore, it must help to place boundaries between exact science and knowledge on one hand and inexact science and knowledge on the other hand. Here, some aspects of philosophy claim that science, generally considered, is a reasoned-based analysis of sensation upon either awareness or acquaintance. From this conceptualization, this particular view further claims that a scientific method cannot deduce anything beyond *what is observable* by the existing or theoretical means. The search for an optimal path for knowledge production is simply a design of an optimal epistemic input-output machine to act on the defective information structure to produce the best output from the defective input.

A number of epistemic questions tend to arise from the cognitive link of observations to knowledge. How exact is what is deduced from the observables and to what extent are the observables related to the elements of the universal object set (the potential or the ontological space)? One thing seems to present itself as universal in the knowledge production process. The universality is that every epistemic item claimed to be knowledge in science and by any scientific method, empirical or formal, is conditional and inexact. Exactness is claimed from inexactness by decision-choice actions that allow cognitive agents to abstract exact-value equivalences from the epistemic constructs. This follows from the basic universal claim that all cognitive agents work with epistemological defective information structure and have no unquestionable claim to the ontological perfect (non-defective) information structure. There is always a core of irreducible inexactness in any area of knowledge production. This core of inexactness constitutes the sum of irreducible cores of stochastic and fuzzy uncertainties that are generated by the epistemological defective information structure. The claim of the qualitative property of exactness as associated with a category of science is subjective, that always has an implied fuzzy-stochastic conditionality which may or may not be explicitly stated. The instrumentations of measurements and observations in all knowledge areas are human created and cannot be claimed to be unquestionably objective and hence exact. We work with vague concepts, inexact units of measurements and approximate reasoning in every area of science over the epistemological space. The presence of accidents and systems' failure in mechanical systems are testimonies to the inexactness of science and other areas of knowledge. This is pointed out by Max Black as:

> *The ideal standard of precision which those have in mind who use vagueness as a term of reproach, when it is not a shifting standard of relatively less vague symbol, is the standard of scientific precision. But the indeterminacy which is characteristic of vagueness is present also in all scientific measurement.*

In support of this statement Black quotes N. R. Campbell as:

> *'There is no experimental method of assigning numericals in a manner which is free from error. If we limit ourselves strictly to experimental facts we recognize that there is no such thing as true measurement, and therefore no such thing as an error involved in departure from it.' Vagueness is a feature of scientific as of other discourse* [R19.4, p. 429].

There are other philosophical claims about naturalism of scientific methods. The claims with scientific methods must adhere to an empirical study and must have an independent verification as well as a process for properly developing and evaluating the natural explanations for observable phenomena. Critical rationalism, as it relates to the property of exactness of science, is that unbiased observation is an impossible task and hence demarcation between natural and supernatural explanation is simply arbitrarily done on the basis of human action in decision. The claim of critical rationalism to support the empirical property of science is that *falsificability* must not be the landmark of empirical theories but that falsification must be the universal empirical method of science. In this way, science must be viewed and assessed in terms of its ability to increase the scope of testable knowledge against accumulated experience.

Interestingly, the position of the critical rationalists fails to acknowledge that every scientific knowledge or any knowledge item has seeds of fallibility and infallibility that exist as *duality in logical unity* which functions under the principle of continuum, and that every test of scientific knowledge that is accepted to meet current accumulated data is conditional, not only on as yet unknown data, but relative to hidden component of falsehood, all of which reside in the *penumbral region* defined by vagueness and ambiguities which require clarification and acceptance by cognitive agents through judicious judgments and decision-choice actions. The process of falsification comes with errors of data, vagueness in representation and errors of human judgment regarding the knowledge-acceptance process in the epistemological space. Over the epistemological space, we cannot overlook the presence of epistemic errors which come to us as a composite sum of errors of fuzziness, errors of information incompleteness and errors of reasoning. It is due to these errors in the epistemological space that we encounter situations where epistemic items which are claimed to contain knowledge are rejected and latter found to be knowledge-wealthy and vice versa. Here, the development of classical mathematics in the probability space provides us with error measuring capacity due to the information incompleteness. The contemporary expansion of this classical mathematics into fuzzy mathematics in both possibility and probability spaces provides us with error-measuring capacity due to information vagueness and incompleteness. Together, all of these generate a knowledge-production system that is internally self-correcting for epistemic progress toward exactness and certainty.

## 7.3  Science, Grammar and Acceptance of Knowledge

The definition of science has gone through and is still going through refinements. Definitions of science in our contemporary thought system seem to suggest that the linguistic category of science has the properties of : 1) formulation of models in natural science; 2) Explanation; 3) Prediction in relation to two structures which are presented as characteristic sets in Table 7.3.1 that must be related to the characteristic sets I, II and III in Tables 7.2.1.11 – 7.2.1.3.

**Table 7.3.1** The Characteristic Set for the Definitional Structure of Contemporary Science

**The Essentials of Definitional Structure I**
a) Process of knowing, b) Knowing is represented by a model c) A model explains the actual (present) and predicts the potential (future), d) The prediction of the future (potential) is a test against the past, and e) Objectivity in the knowledge-construction process is independent of human cognition.

**The Essentials of Definitional Structure II**

f) Controlled methods for data collection by observation of physical evidence of the phenomenon, g) Construction of theoretical explanations of the collected data, h) An experimental and empirical research under methods of generalized hypothesis and i) Tests of hypotheses under controlled conditions.

Given all these suggested characteristics of science and hence its definition as a linguistic category, we may ask the question as to what is the objective of science or knowledge in general? The question of the objective of science pushes us into asking five related questions.

**Table 7.3.2** Questions on the Objective of Science and Knowledge

1. Is the objective (goal) of science the focus on understanding nature by theory and models of explanation with abstracted *explanatory rationality* about *what there is*?

2. Is the objective (goal) of science the focus on offering changes of *what there is* through theory and models of prescription with *prescriptive rationality* in order to improve the general and environmental conditions of human existence?

3. Do the experimental methods used in data-creation process pose problems of empirical rationality in such a way that partial answers are only available to us regarding the phenomenon of interest?

4. Do the instruments used in knowledge experiments to answer stated questions only produce answers to these questions that are meaningful only in terms of the instruments employed?

5. In all these, what role do axiomatic system with derived *axiomatic rationality,* and empirical system with established *empirical rationality* play in the knowledge production in both explanatory and prescriptive thoughts? In other words, what is the relationship between theoretical and empirical sciences? Additionally, is the distinction between empirical and theoretical sciences universal to all areas in the knowledge-production process?

With these properties, and to the extent to which they relate to goals and objectives of science as a category of the knowledge-production process, we may advance a working definition of science. The working definition will allow us to create its linguistic category that will in turn offer us an algorithm of dynamic sorting of science and non-science categories at each point of time. Let us keep in mind that science and non-science dwell in a dynamic unity of duality and continuum where an element of a non-science of today may come to be classified as an element of science tomorrow through a process, as new information reveals new possibilities in the knowledge production over the epistemological space. Science has been projected as a derivative of knowledge, and as such, we need a provision for a working definition of the concept of knowledge before we can proceed to the definition of the concept of science. The definition of knowledge is a categorial conversion of a linguistic category of information which is the connecting element of all the knowledge-production processes.

The characteristics of knowledge as a linguistic category must be specified. The importance of the linguistic category as a framework of definition and analysis must not be underestimated. Its cognitive beauty and relevance for our understanding rest on our knowledge production which proceeds from conditions of ignorance through categorial conversion of linguistic categories to conditions of knowing. The categorial conversion is accomplished by logical operations where there is always a *primary linguistic category* and *derived linguistic categories* proceeding through the stages of the *knowledge square* as we have discussed in the previous chapters. The approach of linguistic category-formation and categorial conversion is taken to be that which corresponds to cognitive actions that deal with epistemic states, processes and time for understanding and knowing. Every categorial conversion comes with a categorial moment and transitional dynamics of quality and quantity among categories with neutrality of time whether such a categorial conversion relates to natural or social phenomenon. The linguistic category is closely related to definitions and conditions of definability, grammar of knowledge as well as to the explications and conditions of explicability. Let us turn our attention to a working definition of knowledge.

### Definition 7.3.1: Knowledge

Knowledge is a set of cognitive properties of epistemic objects with a justified true belief from information signals where the structure of such a justification must be provided to explain how it relates to the subjective information set as the evidential set that provides the justification support of the logical claims of epistemic items as knowledge. The knowledge production is a set of cognitive actions with epistemic processes that link the ontological space to the epistemological space and then to the ontological space by justified claims to ensure a minimal difference between the epistemic and ontological realities.

### Definition 7.3.2: Justification

Justification is an interpreted body of information signals with processed relations to one another that establish a relational structure between a set of linguistic objects and epistemic objects in the universal object set in order to establish a belief system of a set of knowledge objects. It is the set of belief conditions that supports the claimed relationship between epistemic elements and an ontological element by cognitive agents.

As defined, knowledge is a language composed of syntactics, semantics and pragmatics that relate linguistic objects to the elements in the universal object set for understanding and use. Knowledge is a linguistic category that allows us to sort out that which is simply opinions or non-knowledge in the sense that these opinions have no justified belief support. As a category, knowledge constitutes a set of elements. In this respect, human experiences, $\mathbb{H}$, may be said to be composed of knowledge, $\mathbb{K}$, and non-knowledge $\mathbb{K}'$ such that $\mathbb{H} = \mathbb{K} \cup \mathbb{K}'$ with $\mathbb{H} = \mathbb{K} \cap \mathbb{K}' = \varnothing$ in the Aristotelian logical system within dualism operating under the principle of excluded middle. We shall refer to this as the *Aristotelian knowledge category*. The conditions where $\mathbb{H} = \mathbb{K} \cup \mathbb{K}'$ and $\mathbb{H} = \mathbb{K} \cap \mathbb{K}' \neq \varnothing$ hold in the fuzzy logical system within duality where the principle of continuum comes to replace the principle of excluded middle. We shall refer to this as the *fuzzy knowledge category*. The Aristotelian system, in category formation, places human cognitive experiences in either knowledge or no-knowledge exact categories with surety.

The fuzzy logical system, in category formation, places human cognitive experiences in fuzzy categories of knowledge, no-knowledge and unresolved zone where each category has a corresponding distribution of degrees of confidence associated with knowledge and no-knowledge. Every knowledge category is supported by a justification principle, $\mathbb{J}$, that provides its belief support, $\mathbb{B}$. Every knowledge item has two awareness components of individually subjective conditions of awareness, $\mathbb{A}_i$ and collectively subjective conditions of awareness,

$$\mathbb{A}_C = \bigcup_i \mathbb{A}_i \text{ with } \bigcap_i \mathbb{A}_i \neq \varnothing \text{ in the sense that some of the individual awareness}$$

conditions may be overlapping. The claim of any social knowledge is a combination of awareness condition and justification principle that define a belief system in support of a claim that an epistemic item contains knowledge. As specified, there is a set of epistemic elements with a corresponding set of justification that corresponds to an ontological element. The corresponding justifications constitute theories and models that must be tested against the correspondence to the an ontological element in the sense that the epistemic characteristics correspond to the ontological characteristics where the acceptance of the an epistemic item as containing knowledge is a decision-choice action with the highest degree of belief as can be ascertained.

As it has been argued in [R2.9] that our knowledge-production system derives its growth and development from transformation dynamics of subjective–objective duality that resides in the information polarity, where information on ontological reality is under continual categorial conversions to subjective information and the subjective information is under categorial conversions into elements of epistemic reality in the epistemological space. Alternatively stated, the knowledge dynamics is such that the ontological information is transformed into the epistemological information by a process. The epistemological information is then transformed into epistemic reality. The transitional cognitive dynamics of the knowledge-production process requires a *rational justification* which is a justified belief that a knowledge item has been obtained; a *rational falsification* that provides a second order belief that the justified   information is reasonable   and a *rational corroboration* that provides a third order belief that a knowledge item is found in the set of items of cognitive reality; and the three rational conditions must be linked by a *rational verification* that provides a combined justified belief that the epistemic item is identical to an item in the ontological reality and hence a knowledge has been found in the search process. This process may be illustrated as *belief* and *analytical squares* that are categorial derivatives from the knowledge square which is taken to be a primary category of knowing that revolves around the space of epistemic reality as illustrated in Figure 7.3.1.

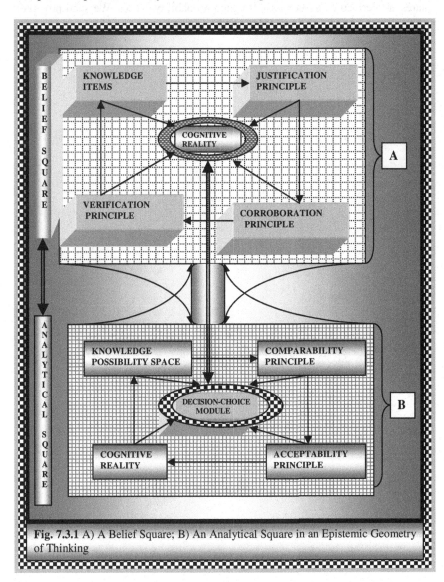

**Fig. 7.3.1** A) A Belief Square; B) An Analytical Square in an Epistemic Geometry of Thinking

The theory of the knowledge square provides us with a universal principle as a meta-theory on epistemology for the knowledge-production process. The *belief square* presents a relational structure among subjective knowledge, justification conditions and derived corroboration conditions to allow verifications of the claimed knowledge items against elements in the space of ontological reality. The *analytical square* is a derived category from the belief square. It is a cognitive relational structure among the elements of the possibility space, where each element must meet the comparability principle and acceptability principle for an entry into epistemic reality space. The conditions for both the belief and analytical squares are derived from the possibility and probability spaces. We shall now turn our attention to the interactions of knowledge and science.

# Chapter 8
# Knowledge and Science in the Theory of the Knowledge Square

Science is a sub-category of knowledge and hence it must satisfy the general conditions of the primary elements of the knowledge square and its derivatives of belief and analytical squares. As a sub-category of knowledge, what distinguishes it from other sub-categories of knowledge? This question demands us to produce a definition of science that will allow us to create its linguistic dynamic category which will offer us the conditions of a sorting algorithm for locating scientific items and non-scientific items at any point of time in duality. We have already offered a definition of knowledge. We shall now offer a definition of science as a sub-category of knowledge.

## 8.1 The Identities of Knowledge and Science

It has been argued that every epistemic element or an ontological element has an identity that allows it to be identified. The identity is revealed by a set of characteristics that allows us to place the elements in their respective categories. The identity of the ontological element is revealed by the ontological characteristic set in terms of *what there is*. The identity of an epistemic element is revealed by an epistemic characteristic set in terms of what is claimed to be acquainted or known. The linguistic category allows the cognitive construct of nominalism, that itself leads to a methodological constructionism and reductionism in the epistemological space for knowledge development to create the epistemic reality that must be examined against the ontological reality. The comparison and reconciliation of the characteristics between epistemic reality and ontological reality leads to claims of science and non-science.

**Definition 8.1.1: Science**

Science is a knowledge sub-category that is defined by a dynamic set of epistemic properties which allows the conditions of dynamic sorting algorithms to be developed in order to establish a demarcation between science and other categories of knowledge as we travel through time.

K.K. Dompere: The Theory of the Knowledge Square, STUDFUZZ 289, pp. 139–161.
springerlink.com                                      © Springer-Verlag Berlin Heidelberg 2013

The definition of science is intentionally broad for flexible accommodation and subject-quality dynamics of different areas of knowledge. As defined, science cannot be explicated by the content of subject areas of knowledge. Every subject area of knowledge is a candidate to be placed in the linguistic category of science, as new and powerful epistemic tools are cognitively uncovered in such a way that the knowledge areas may be examined against the specified characteristic of the linguistic category of science. If contents cannot be used to define the knowledge area as science, then what characteristics should be enlisted to establish the boundaries of science as a linguistic sub-category of knowledge? This question is troubling; however, we must look not to the content of the knowledge area, but to the methods of obtaining and accepting a particular item as knowledge in order to establish the domain of science. It has been pointed out that the theory of knowledge square with its belief and analytical squares provides a *universal principle* for the knowledge-construction and knowledge-reduction processes irrespective of the area and subject matter of interest. The universal principle simply states that all knowledge productions take place in the epistemological space and that they are the works of information and decision-choice rationality of cognitive agents. Additionally, every information structure in the epistemological space, as an input into the knowledge-production process is defective, and its decision-choice rationality is fuzzy irrespective of the subject area. The fuzzy rationality is an enveloping of the classical rationality which is its point-point approximations. The methods of epistemic activities will be an enveloping of the sequential paths from conditions of cognitive ignorance to a claim and acceptance of knowledge. The sequential path and the methodology of knowledge acquisition must be explicit and outlined for general discussions and understanding. Here, as we have argued in this monograph, *methodology* is distinguished from *epistemology* while their combined structure constitutes the *theory of knowledge*.

When we accept methods and techniques as candidates in defining the linguistic category of science in the enterprise of knowledge production, then a set of questions comes to the surface for critical deliberation. The set includes: a) What role do theoretical or axiomatic methods play in explication of science? b) What role do empirical methods play in establishing the boundaries of science? c) What relationships exist between theoretical and empirical methods in the explication of science? These questions challenge us to distinguish empirical methods from theoretical methods and their interpretive roles in defining science. In this distinction, an empirical theory has been introduced as a distinct theory from axiomatic theory as a system of methods that presents science as a linguistic category.

Here, we may choose to speak of an empirical representation of ideas of *what there is* as an ontological description which is an epistemic representation of ontological elements and an empirical explanation of *what there is* and its behavior as an epistemic understanding of an ontological description. The former is the *empirical ontology* as sense data and the latter is the *empirical epistemology* in the knowledge-production process to reduce human ignorance. The empirical ontology presents us with information, on the basis of experience and observations, as input for processing while the empirical epistemology presents us with a template of information processing capabilities on the basis of empirical methods. We may also choose to speak of an axiomatic representation of ideas of *what there is* as an ontological description and an axiomatic explanation of *what there is* and its behavior as an epistemic understanding of the ontological description. The former is an *axiomatic ontology* and the latter is an *axiomatic epistemology* in the knowledge-production process. Both the axiomatic and empirical representations of thought are nothing but reflections or models of sense data that one uses to interpret information signals from the ontological space [R2.9] [R8.18][R8.36]. The empirical and axiomatic representations of experiential information become inputs into the cognitive actions that require reasoning (logic) to process them into thoughts, thus establishing connections between ontology and epistemology. The relational structure of specifying the primary category and derived categories in knowing is presented as a cognitive geometry in Figure 8.1.1.

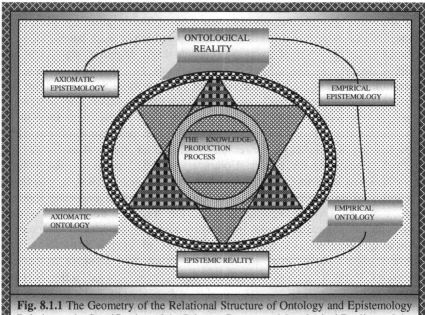

**Fig. 8.1.1** The Geometry of the Relational Structure of Ontology and Epistemology Relative to the Specification of the Primary Category of Ontological Reality and the process of obtaining the Derived Categories of Epistemic Reality

The knowledge-production process that generates the ontological-epistemological relational structure may be linked to the nature of the information structure for the initialization of the knowledge search. The axiomatic or the empirical information structure may be taken to be fully defective (both vague and limited information), partially defective (either full but vague or non-vague and limited information) or non-defective (perfect information: non-vague and full information).

One may ask a question as to what are the differences and similarities between the *empirical ontology* and the *axiomatic ontology*. Similarly, a question may be raised as to the differences and similarities between the empirical and axiomatic epistemologies. At the level of empirical ontology, direct sense data is its domain of characteristic representation to generate the information inputs that present identities of epistemic objects in the epistemological space for epistemic processing. This requires an acquaintance with *what there is* as defining the primary category of knowing *what there is*. At the level of axiomatic ontology, cognitive intuition in communications provides its domain of characteristic representation of abstract properties of *what there is* that will constitute the information inputs in the epistemological space for epistemic processing. At the level of empirical epistemology, the cognitive algorithms are developed to process the empirical characteristics of *what there is* to obtain a thought of explanation of either *what there is* or the behavior of *what there is*. This requires empirical laws of thought that may be based on either classical paradigm or fuzzy paradigm.

Similarly, at the level of axiomatic epistemology, cognitive algorithms are developed to process the axiomatic characteristics of *what there is* into the explanation of *what there is* or the behavior of *what there is* as seen from the internal structure of the axiomatic system. This requires axiomatic laws of thought that may also be based on either classical paradigm or fuzzy paradigm. In all these cases, decision-choice rationality is required. The processing of empirical representations of information by reason to achieve an empirical explanation in thought by combining the empirical ontology with the empirical epistemology may be called an *empirical theory* that presents an explanatory view of *what there is* in terms of existence and behavior. Thus, an empirical reasoning module becomes the *cognitive categorial moment*, the force of transformation that converts the ontological categories into epistemological categories. In the process of knowing, therefore, we have two categories of empirical ontology and empirical epistemology that are linked together by cognitive categorial conversions through epistemic processes and decision-choice actions. Here, ontology and epistemology exist as a duality in logical unity and continuum but not as dualism in logical separation and excluded middle of the knowledge-production process. The empirical ontology provides the conditions to establish the primary category while the empirical epistemology creates the conditions to construct derived categories for epistemic reality.

An axiomatic representation in thought is also nothing but a reflection of the properties of that which exists as ontological element. It may or may not flow from sense data as representation of information signals taken as self-evident truth. In this approach, the axiomatic representation of ontological characteristics becomes an input into cognitive actions that require a further reasoning to process it into complete thought. The process presents *axiomatic ontology* and *axiomatic epistemology* as connected by categorial conversion in duality in continuum which becomes the dynamic carrier of reasoning that provides the logical energy of categorial conversion over the transient process from ontology to epistemology. The processing of axiomatic representation in thought by reason to achieve an axiomatic explanation in thought may be called *axiomatic theory*, that presents an explanatory view of *what there is* in terms of existence and behavior. The axiomatic ontology, like the empirical ontology, provides the needed conditions to establish the primary category while the axiomatic epistemology creates the conditions to construct the derived categories of the epistemic reality. Both the empirical and axiomatic conditions may be used in the production of prescriptive theories in prescriptive science.

The knowledge-production process that generates the ontological-epistemological relational structure as seen in terms of axiomatic and empirical foundations in Figure 8.1.1 may be linked to the possible assumptions of the nature of the information structure for the initialization of the knowledge search and to the paradigms for the epistemic processes. The axiomatic or the empirical information structure that initializes the primary category as an input may be takento be fully defective (both vague and limited information), partially defective (either full but vague or non-vague and limited information) or non-defective (perfect information: non-vague and full information). On the basis of the assumed information structure, symbolic representations are developed. From the nature of symbolic representations paradigms are formed and advanced for thought processing. To present the relevant information assumptions, we shall take up the relational structure of information and ontology on one hand, and then information and epistemology on the other hand.

The epistemic algorithms constitute the laws of thought that are made up of two steps. The first step involves the principle of reasoning to define the basic transformation characteristics required to convert ontology to epistemology. The second step involves the principle of reasoning that determines which of the characteristics constitute legitimate derivatives in the epistemology. Here, we must distinguish the empirical laws of thought from the axiomatic laws of thought in any paradigm even though both of them constitute cognitive conversion moments from the ontological categories to the epistemological categories. All these are connected to cognitive transformations of knowing but say nothing about the truth of the primary and derived categories and the propositions they contain. It will become clear that some area of the knowledge production lend themselves to both empirical ontology and epistemology, some lend themselves to both axiomatic ontology and epistemology, while other areas will require a combination principle of both.

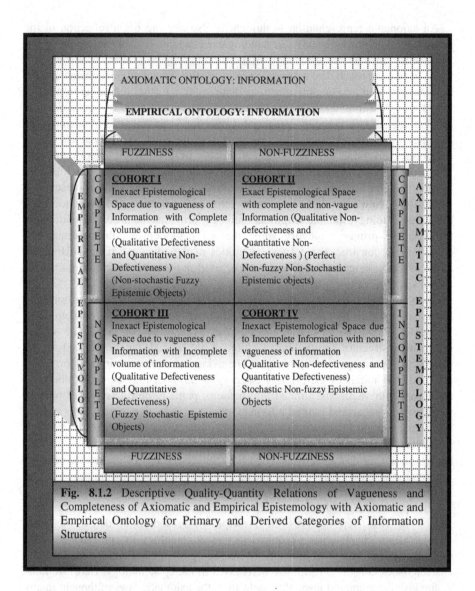

**Fig. 8.1.2** Descriptive Quality-Quantity Relations of Vagueness and Completeness of Axiomatic and Empirical Epistemology with Axiomatic and Empirical Ontology for Primary and Derived Categories of Information Structures

## 8.2  Objectives and Classification of Science: The Inexactness of Exact Science

While it is not difficult to make an argument that the classification of science must follow the empirical explication of science, an argument in support of it is not easy. The problem is whether the goals and objectives of science should be among the explication's characteristics of science or should science be explicated before we establish its goals and objectives. The importance of understanding the nature of the set of goals and objectives of science will become clear when we come to deal with the problem of vagueness in the content and methods of science. We shall now deal with possible goals and objective of science.

## 8.2.1   Goals and Objectives of Science

In conceptualizing the goals and objectives of science, there are few questions, the answers of which will help our concept formations, given the explication of science. The set of questions may be placed under two classifications: one at the level of abstraction and theory, Table 8.2.1.1, and the other, at the level of experimentation and empirical work, Table 8.2.1.2.

---

**Table 8.2.1.1 At the level of abstraction and theory:**

**a)** Is the objective of science the focus on understanding nature through an explanatory theory or is it about the discovery of *what there is* or both? In other words, is science about finding out *what there is* and then explaining the behavior of *what there is*? Alternatively, is the focus of science about ontology or epistemology or both?

**b)** Is the objective of science the focus on prescription in order to change "what there is" so as to improve the quality of nature and conditions of human existence as humanly decided through a prescriptive theory?

**c)** In what framework is the prescriptive process connected to either ontology or epistemology or both?

**d)** Can we claim *exactness* of abstraction and theoretical results and under what set of conditions?

---

### Table 8.2.1.2 At the level of experimentation and empirical work:

a) Do the instruments used in experimental methods or knowledge experiments produce exact or vague answers to the questions that are only meaningful in terms of the instrument used? In other words, do experimental processes generate vague data relative to ontological elements?

b) Do the experimental methods used in obtaining empirical data about a phenomenon pose problems such that partial answers are only possible to us at the level of *what there is* and that such partial answers are due to inexactness of the experimental methods?

c) Are experimenters interacting with the problem during the periods of experimental design, recording of the results, collection of the data and analysis of the data? In other words, what role does subjective-objective duality plays in the experimental knowledge where the recorded data may not reflect the experimental outcomes?

d) Can we claim exactness for any of these stages in the process of knowledge production through experimentation and empirical work?

When science is taken to deal with the discovery of *what there is* and the explanation of the behavior of *what there is*, then the ontology precedes the epistemology in the sense that *what there is* must first be anticipated, discovered or conceptualized before the justification and explanation of the behavior of *what there is* become of cognitive interest. In this way, an ontological element is first transformed into an epistemic element and moved to the possibility space. This approach is consistent with all areas of the knowledge-production process but not restricted to science. It is also consistent with the *exact* and *inexact* knowledge-production. The thinking system that follows this path of the knowledge-acquisition process, is called the *explanatory epistemic process* where ontology precedes epistemology, and the area of science is called the *explanatory science* while the corresponding theories are called *explanatory theories* (see also [R14.11] [R14.13] [R14.35] [R14.97] [R14.102]). On the other hand, when science is taken to deal with changing *what there is* and prescribing *what ought to be* for implementations, then epistemology precedes ontology in the sense that *what ought to be* is first anticipated and conceptualized, and the outcome of its prescriptive rules after implementations becomes *what there is*. Similarly, this approach is consistent with the general cognitive processes in examining inexact and exact knowledge production.

In this analytical structure, what do we mean by epistemology precedes ontology? The epistemic process is viewed as a cognitive map (a blueprint) of abstract understanding with a justified principle that allows decision-choice algorithms to be constructed and imposed on human actions to create categorial conversions that form conditions of ontology to transform *what ought to be* into *what there is*. What ought to be is conceptualized in the possibility space and the cognitive element belongs to both the possibility space and the possible-world space. Here, the epistemic element under the prescriptive science belongs to the possibility space if it is supported by a possibilistic belief system. If, however, the element is not supported by a possibilistic belief system, then it belongs to the possible-world space but not to the possibility space. The actualization of *what ought to be,* into *what there is,* is defined by ontological conditions. The characterization of *what ought to be* and the structure of the developed algorithms (optimal decision rules) for its actualization into *what there is,* is defined by epistemological conditions. The thinking system that follows this path of the knowledge-production process is called the *prescriptive epistemic process* where epistemology precedes ontology and the area of science is called the *prescriptive science* while the corresponding theories are called *prescriptive theories.* Examples in practice of prescriptive science are space exploration, socioeconomic planning, engineering of all kinds, some aspects of medical sciences and bioengineering, genetic modifications, artistic creations and many forms that are emerging. The geometry of the explanatory-prescriptive relational structure for cognitive activities over the epistemological space is provided in Figure.

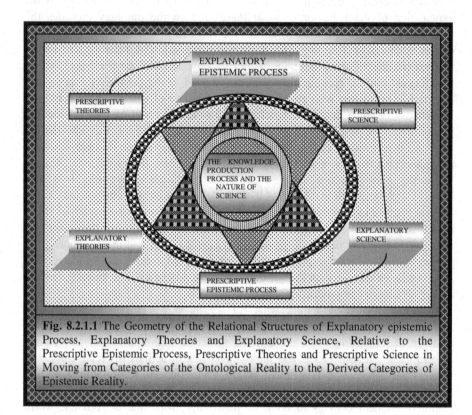

**Fig. 8.2.1.1** The Geometry of the Relational Structures of Explanatory epistemic Process, Explanatory Theories and Explanatory Science, Relative to the Prescriptive Epistemic Process, Prescriptive Theories and Prescriptive Science in Moving from Categories of the Ontological Reality to the Derived Categories of Epistemic Reality.

## 8.2.2  Classification of Science

Given the defined conditions of science, the space of sciences has been and continues to be sub-classified into sub-categories. These sub-categories have their own defined characteristics. Some of the sub-categories are based on areas of knowledge, methods of research and knowledge acquisition, and the content of research and knowledge. The basic structure of such a classification is presented in Figure 8.2.2.1. It allows us to place under a proper epistemic position the classical and fuzzy paradigms with relative understanding of the relative roles played by the *law of excluded middle* and the *law of fuzzy continuum* in the knowledge-production process as we encounter conditions of inexactness, vagueness, quality and quantity.

In these sub-classifications of science, we must pay a special attention to scientific activities and the whole structure of the knowledge-production enterprise. Formal science is grounded on the basis of axioms from which

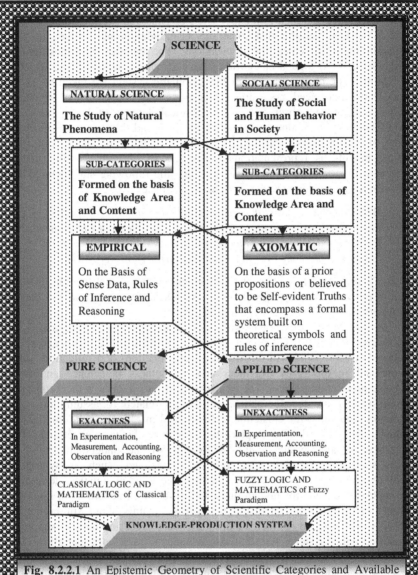

**Fig. 8.2.2.1** An Epistemic Geometry of Scientific Categories and Available Logics

theoretical ideas are then developed within the axiomatic system by an acceptable mode of reasoning to erect a theoretical system of propositions with truth values that are dependent on the truth values of the basic axioms and the method of thought. These are the conditions that allow mathematics to be placed in the category of science and then into exact science. Thus, the starting point of formal science is the basic axioms that constitute its foundation and capture the initial information input for categorial conversions. The theoretical propositions are hierarchy of derived logical categories whose primary category is the set of the basic axioms. The basic elements of the axiomatic system are taken to be self-evident truths that need no empirical justifications. Experimental science is viewed as starting from observations of ontological elements of the *real world* in a defined laboratory to accumulate experiential data, and through some acceptable methods of reasoning, one derives, on the basis of the accumulated data, useful empirical models as representations of aspects of epistemic reality that are consistent with the observations. It is here that the following statement by Max Black acquires an epistemic potency in the fuzzy space.

> *While the mathematician constructs a theory in terms of 'perfect objects', the experimental scientist observes objects of which the properties demanded by theory are and can, in the very nature of measurement, be only approximately true.* [R19.4, p. 427]

It is also, here, that the concept of exact science without qualification looses its epistemic relevance and the problem of exact science enters into the class of phantom problems.

Our discussion, here, is not to visit the old debate between empiricism and metaphysical rationalism. The objective here is to reexamine thoughts and present a relational structure that will allow critical examinations and reflections on the concepts and meanings of exactness and inexactness in science and in the knowledge-production process. The objective is also to show how their conceptual meanings reflect on decision-choice actions as related to assumptions about information inputs into the epistemic technological machine and the implied laws of thought. This will allow us to place the differences between exact and inexact sciences on an appropriate analytical footing for the understanding of new sciences such as systemicity and the emerging ones, such as synergetics, complexity theory energetics, super-symetric theory, adinkramatics, adinkralogy and complexity science for human quest for increasing freedom from ignorance. We hope to arrive at a conclusion that certain claims about similarities and differences of exact and inexact sciences are ideological and hard to support by any justified belief system. We shall also advance a framework that allows us to reject the notion of *absolute truth* in knowledge when we retain the idea that knowledge is a model of reality constructed from an inexact epistemological space.

From the structure of the theory of the *knowledge square,* what we know as characterizing reality is a cognitive mapping from the universal object set (the universal space) onto the space of epistemic reality, thus producing epistemic objects with belief support systems. Let us keep in mind that metaphysics means everything that is non-experiential and comes to us as rationalism. Mathematics and symbolic logic are always associated with axiomatic science and hence rationalism. Empiricism is connected to everything experiential. There is a complete marriage between methodological empiricism and methodological rationalism in the search for knowledge that is accepted as scientific. The two are needed for cognitive activities in the totality of the knowledge-production process. The needed examination has been made necessary because of an increasing development of the fuzzy paradigm in the knowledge-production process. This fuzzy paradigm is not limited to any specific area of knowledge production and its relevance may be seen in terms of our ability to process defective information structure on the path of knowing.

The characteristics which distinguish the linguistic category of science from that of non-science within the organic category of knowledge in the epistemological space seem to be nothing but the methods of knowledge acquisition where methodological empiricism and methodological rationalism are inter-supportive. We have shown from the theory of the knowledge square that one of the universal principles for the knowledge-production process, irrespective of the knowledge area, is that of the *principle of defective information structure* where every knowledge area must deal with the conditions of inexactness. The unifying force is the *information signals* that are transmitted from the ontological space to the epistemological space in order to help to link experiential factors to elements of formal factors. In this respect, *a-prior* and *a-posterior* propositions are merely matters of methodological convenience that depend on the subject matter of an analytical interest. In the rationalist-empiricist approach, a preference is offered to the investigator in terms of an initial choice of methodological approach where rationalism or empiricism may be taken as the starting point of an inquiry. The choice will depend on the subject area of the knowledge production, things that are taken (assumptions) to initialize the epistemic journey and the vehicle that will take us on the path to the house of knowledge. And even, here, we are confronted with an awesome problem of dealing with the initial problem of inexactness.

The choice of the initial specification as either exact or inexact information structure is similar to the choice between *a-prior* propositions and *a-posterior* propositions, and also similarly, to the choice between prescriptive propositions and explanatory propositions as initializing the knowledge-production process. In these discussions, prescriptive propositions flow from the structure of prescriptive theory and science as have been presented and discussed in this volume and also in [R14.30] [R16.4] [R16.7]. These prescriptive propositions are not derived from explanatory theory. They are derived from an axiomatic system. In this respect, it is useful to distinguish between *real variables* that may represent *real objects* and *linguistic variables* that may represent *linguistic objects*. The objective of the knowledge-production process is to establish relationships among real objects and linguistic objects in order to claim knowledge through a process. The real objects are not under the domain of human existence and action but rather ontologically given. The linguistic objects are under the domain of human existence and action, and hence require information processing through cognitive algorithms to abstract and test their correspondence with the real variables in the universal object set. In other words, in the knowledge-production enterprise and scientific discovery, we take the set of the real objects, processes and states as given and as the primary category of reality with linguistic descriptions that represents the potential space of knowing.

This potential space is the ontological space that contains the universal object set with cardinality equal to infinity, and is also infinitely closed under natural substitution-transformation process without loss of matter and energy, but with losses of quality which are compensated by other gains of quality that relate to information signals. Here, we proceed on the principles of epistemic constructionism where our linguistic objects constitute the derived category from the primary category, and where the instruments of transformation are cognitive algorithms that create logical conversion moments that help to define qualitative and quantitative motions between ignorance and knowledge with neutrality of time. The epistemic problem of the knowledge construction, therefore, is how to create a *linguistic world* whose concepts, that define the derived categories, might be abstracted from the primitive concepts that correspond to the real objects of the primary category. Alternatively, the problem may be seen from the dual in terms of methodological reductionism in the knowledge creation in terms of how to take the derived categories, massage them and completely reduce them to correspond to the primitive concepts that relate to the real objects of the primary category of reality. In this process of knowing, empirical science finds a distinction from axiomatic science as well as a point of unity under conditions of categorial conversion, relationality of thought and logical continuum. As argued, ontological reality constitutes the primary category of the knowledge production, while the epistemic reality constitutes the derive category of knowledge.

At the level of empirical methods of knowledge finding and construction, we initialize the knowledge production with a set of empirical propositions that has a *parallel correspondence* to the items of the primary category (universal object set). If we accept the empirical inferential rules that are taken to reflect or represent the objective laws of the primary category, and if the initialized set of empirical propositions corresponds to the set of conditions of the primary category, then we may come to believe that the derived empirical propositions are consistent with the primary category of the universal object set as empirically established and with respect to the empirical logical rules. Under these conditions, we can claim a knowledge item with justified belief conditions where such empirical propositions are provable within the empirical system of thought. Here, we speak of *empirical completeness*. The concept of parallelism must be interpreted as consistent and similar. How exact are our empirical methods of inference and the derived propositions; and hence how exact is the science that employs such methods given the ontological exactness?

At the level of axiomatic methods of the knowledge construction, one initializes the path to knowledge discovery or knowledge search with a set of axiomatic propositions that has a correspondence with the elements of the primary category. If we accept the use of axiomatic logical inferential rules that provide us with a justified reflection of the objective laws of the existence of the primary category, and if the initial set of axioms is shown to correspond to the set of conditions of the primary category, then we can construct a belief support for the axiomatic claim of knowledge. In this respect, the derived axiomatic propositions are consistent with the elements of the primary category of reality. In this connection, we speak of *axiomatic completeness* in the sense that the axiomatic system is closed to allow us to justify the claims as knowledge within that system of thought. How exact are the axiomatic methods of inference and the derived propositions, and hence how exact is the science that employs such methods given the ontological exactness?

As it stands, empirical and axiomatic methods defer from their initialization of the road to the knowledge production in terms of what is taken to characterize the elements in the primary category of logical reality and the logical rules for obtaining the derived category and the knowledge acceptance conditions. Given the selected primary category, the process of obtaining the derived categories, at each stage of knowing, is governed by a language and rules of combination of terms that involve synthetic and semantic theories of the behavior of the elements of the linguistic categories in any given language. The words in the language constitute different colors with differential shades of meaning that must be carefully selected in combination to paint the picture of the claimed truths. As colors, words present grey areas of semantics and their syntactic construction cannot be claimed to lead to unquestionable conclusions even when the primary category is accorded logical legitimacy.

Science or knowledge in general, viewed either from the empirical or axiomatic system, presents itself as a category of spoken words or written characters or symbolic representations or both. It is here that empirical system finds unity with axiomatic system and methodological positivism finds comfort with philosophical rationalism. It is also here that exact and inexact sciences may be shown to have an inseparable unity with a common ancestry. It is also here that a distinction between formal science and experimental science must be viewed as a choice of convenience of linguistic category that must deal with the conditions of inexactness. Formal science proceeds with an initialized theoretical system and conceptual ideas as a description of the primary category that leads to derived categories of a theoretical system by construct. Experimental science, on the other hand, proceeds with initialized observational data as the description of the primary category that leads to derived categories of an experiential system by construct.

On the question of claims to knowledge, one may hold a position that nothing can be learned about reality from studying formal sciences such as mathematics. Similarly, one may hold a position that nothing can be proven to be a reality through the use of formal science, and that applied mathematics is a utilization of theoretical and mathematical models simply to understand epistemic reality within limits. In this connection, are symbolic logic and mathematical logic categories of syntax or science of reason? How do they influence or affect the results of empirical and axiomatic sciences? The disciplines of formal sciences or axiomatic systems rely heavily on the use of mathematics. They, therefore, do not exist until favorable mathematical forms are developed into advanced systems. These questions lead to an important question as to how do empirical, axiomatic and mathematical sciences deal with the difficulties of processing defective information structure. In fact, this is a major problem in theoretical economics and mathematical economics where human decision-choice behaviors in the economic space are axiomatized into the primary logical reality from which derived epistemic realities are constructed in terms of hypothesis and propositions. It is also a problem in biology, mathematical bioscience, medical sciences, psychology, mathematical psychology, mathematical social sciences and others. The problem is that all these are living systems and behave differently from physical systems. The differences reflect themselves in qualitative dispositions with changing characters through categories of internal qualitative self-motions.

The users of the classical paradigm with its mathematics and logic in the study of non-physical (living) systems have phantom imagination, such that they can arrange the mathematical and symbolic epistemic objects of the living systems with rigid rules just like those epistemic objects in the chess-board of physical systems, with the same great ease as the chess player does on the board of chess elements or draft elements. The users of the classical paradigm fail to account for the presence of the fundamental principle of internal motions, in that the elements on the epistemic game of the chess-board of the physical system retain the same qualitative disposition and do not produce counter motion in response. The elements on the chess-board, therefore, have easy representations and manipulations by the users of the classical paradigm in the chess-board of the physical system. In practice, this conceptual system alters its form to incorporate quality and quantity in the epistemic chess-board of living systems where every epistemic object has imbued in it the principle of qualitative self-motion given the quantity. The epistemic objects of living systems present different challenges to the mathematical and logical representation and processing of information by the logical forces of the classical paradigm.

The above discussions relate to the questions surrounding the *language* and *grammar* of science and mathematics in terms of exactness and clarity of representation and reason. All intelligible scientific and mathematical states, no matter how they are viewed, have a language and a grammar, and these are driven by concepts of sets, categories, groups and memberships in terms of information characteristics that allow for a partitioning, naming of elements and connectability of transformations of elements in the epistemological space. The restriction on the exactness of statements to be pure in science or knowledge is imposed by initial claims of either exactness or inexactness of the conceptual symbols, syntax and semantics that contained in the knowledge system. The reason is that the path to the knowledge acquisition or scientific practice is a varying enveloping not of a fixed set of clearly defined steps that one can follow to find knowledge. It is a path that passes through two types of uncertainties of *fuzziness* (possibilistic uncertainty) and *limited information* (stochastic uncertainty) that are contained in the defective information structure which makes the surface of the path very slippery.

The journey over the path, therefore, must be carefully navigated by the knowledge researchers between two sets of characteristics of truth and falsity under cognitive tension leading to logical substitution-transformations, where information signals are transformed into knowledge, and knowledge is substituted for ignorance; and where decision-choice actions are accorded the categorial moment in information-knowledge dynamics under an epistemic belief system. There are many paths to deciding on truth-falsity dominance depending on the knowledge area, but the method of deciding has a common principle which we have argued to be captured by *cost-benefit rationality*. The cost-benefit rationality is related to fuzzy-stochastic rationalities of the defective information structure that constrains conditions of exactness and inexactness. The question that arises is: what are the appropriate conditions of methodology that must justify the use of a particular type of methodology in processing the available defective information structure? We turn our attention to this question.

## 8.3   The Scientific and Non-Scientific Methodology

With the discussions on the classification of science within the knowledge structure, let us look at the relative merits of categories of science and non-science as may be seen from the methodology of knowing. The scientific methodology or non-scientific methodology is an epistemic framework of reasoning or thinking within the acceptable boundaries of rational thought in the knowledge acquisition under the conditions of limitation and vagueness of information. What then distinguishes natural science from social science, from medical science, from mathematics and symbolic logic and from others? The view being presented here, is simply that there is an intellectual problem on the part of the tradition to distinguish science from non-science, simply on the conditions of methods and techniques in cognitive actions to determine the epistemic objects. Similarly, the distinction between exact and inexact sciences is restrictive in the process of knowledge construction given the continual expansion of knowledge and technological spaces. This restriction finds expression in the structure of the classical paradigm composed of its logic and mathematics, where vagueness is expunged out of words, syntax, semantics and even pragmatism, that makes symbolic and mathematical logic devoid of quality and qualitative motion. The distinguishing characteristics of science from non-science in accord with the classical thought include:

Here, the progress of knowledge, and hence science, requires the insistence on the unity of data and theory in an inseparable whole in a unified setting. Empirical observations and data provide inputs as well as point to the direction of a theory construction and the type of theory needed. A theory on axiomatic basis may point to data requirements for its verification or falsification or corroboration and application. It is here, that one may find a unified epistemic beauty between the empirical approach to cognition and the axiomatic approach to cognition for both the explanatory and prescriptive sciences. The empirical approach may point to the direction and the needed explanatory theory while the axiomatic approach may point to the required prescriptive theory and the needed empirical observation and data requirements. It is here, also, that we must take careful and judicious steps in

**Distinguishing Characteristics of Science from Non-science**

a) Observation of phenomenon of interest;

b) Observation must be measurable as well as produce accurate data;

c) Observation must also be experimental and the corresponding

experiments must be double blind in method when possible in the use

of samples;

d) The underlining behavior of the phenomenon should have a theory

with a derived hypothesis that is based on a logical reasoning;

e) The logical reasoning must follow some criteria of inference,

demonstration and common sense;

f) The specified propositions of the implied theory must meet the

conditions of evidential test as how they help in the understanding of

the behavior of the phenomenon of interest through explanation and

prediction.

defining the concept of observation, and the data requirements relevant to the classical explanatory theory and the modern prescriptive theory that is independent of explanatory theories. It is also here, that a distinction must be established between the explanatory science and the prescriptive science and how both of them relate to the empirical and axiomatic sciences, which then allows us to place the roles of observation and data requirements in proper perspectives in the search for knowledge and epistemic truth. All these categories of sciences, theories and data cannot be claimed to be unquestionably exact because their inputs are defective information for cognitive actions.

At the level of explanatory theory, in the framework of explanatory science, the data requirements are *ex-ante* of the theory, where the objective of theoretical constructs is to justify *what there is*, or the observed, as well as to explain its behavior under the constructed explanatory rationality given the accepted information representation. The explanatory theory implies prediction and possible prescription ex-post of its construction. The empirical methodology is its principal framework that may be supported by axiomatic methodology for strength and completeness within logical bounds. At the level of prescriptive science, the data requirements are *ex-post* of the theory. Here, the objective of theoretical construct is to design a system of consistent optimal decision rules as a blueprint for implementation, in accordance with the constructed prescriptive theory to change *what there is* to *what ought to be* or to improve *what there is* under the constructed prescriptive rationality. Prescriptive theory implies explanation and prediction *ex-post* of its implementation. The axiomatic methodology is its principal framework which may be supported by an empirical methodology for strength and completeness within logical bounds.

Some care must be exercised in using the techniques and methods of investigation and knowledge search to define science and non-science in our modern knowledge construction at the face of virtual systems. We must resist the easy way of speaking of scientific methods as if the character of science has been explicated, unchanging, understood and has a universal agreement in the critical works of the knowledge-production enterprise. The temptation must be avoided in claiming the value-neutral domain of science which is always advanced from the general conditions of subjectivity, politics, economics, law and the stage of social organization, the level of its knowledge and technical know-how. We must also resist the temptation to claim that scientific knowledge is unbiased and non-political in explanation and problem solving. How can science be subjectively neutral or value-free while it is cognitively dependent on the personality, judgment, social institutions and cognition which are all subjectively entrenched on the knowledge agents and the knowledge-production system? The role of ideology in thought is essential in knowledge stability as well as its progress through its methods of knowledge research, approval and acceptance [R2.9] [R7.13] [R7.14] [R7.17][R7.26] [7.30] [R12.8] [R12.18] [R14.63] [R14.64] [R14.68]. The essentiality of social ideology lies in the notion that ideology may enhance or retard the efficiency of the knowledge-production system.

To claim a value and subjective neutrality in science is to claim an ideological neutrality in human social existence, and hence fail to understand that ideology gives content to human social stability and establishes the conversion moments that propel social progress including science. The value and subjective neutrality in any area of knowledge construction externalizes the knowledge seeker, and such externalization is not consistent with the history and sociology of knowledge where the knowledge-production system is self-exiting, self-correcting and self-containing as a nested set. The value-free and subjectivity-free areas of knowledge are purely ideological claims that also contaminate methods of the knowledge-production process. They have no permanency in cognition. The cognitive permanence of the framework of knowledge construction is uncertainty and subjectivity, where uncertainty relates to *vagueness (fuzziness)* and *information limitations (stochasticity)* and *subjectivity* relates to judgment in human decision-choice actions operating with defective information structure in every aspect of human existence in the epistemological space. The statement holds for science and non-science, exact science and inexact science. To claim exactness for science, one must be able to rid cognition of subjectivity and uncertainty of the two kinds in the epistemological space.

In this respect, and for the progress of common intellectual heritage of humanity, we must resist the tyranny of science as well as the entrenched myths about science in discovering reality where such myths help exact science to maintain its ideological position in society as pure. For example, the statement that claims reality of science to be a physical universe defined by determinable properties is subjective and cannot meet the test conditions of objectivity and exactness as defined by exact science. The claim is, at best, established by decision-choice process at the domain of human thought, and hence operates in the penumbral regions of vagueness, ambiguities, subjectivity and randomness in the process of knowing. We must further refrain from an easy acceptance of the notion that the available methods of science are reliable because they are systematic and methodological that are guided by exact language, exact laws of thought, exact reasoning and exact logic of inference. Is the mathematical theory of personal probability, with its axiomatic foundation, subjective or objective, and, is its use in scientific work exact and value-free? The point here is that there is a degree of inexactness in exact science just as there are distributions of degrees of inexactness and exactness over all knowledge sectors. There are also degrees of exactness in inexact science.

As we have pointed out earlier in the discussions on definition and explication, inexactness and exactness constitute a linguistic duality in the space of meanings and interpretations. Let $\mathbb{E}$ represent the term exactness in the language $\mathbb{L}$ and $\mho$ be the space of meanings with $\mathbb{F}$ as a meaning assignment function that takes $\mho$ onto $\mathbb{E}$ to associate meaning to $\mathbb{E}$ in language $\mathbb{L}$ ; then a set of meaning is formed around $\mathbb{E}$ that may be written as $\mathbb{M} = \left\{ \mathbb{E} = F(m) \mid m \in \mho \right\}$ where $\#\mathbb{M} \geq 1$ with a corresponding fuzzy set $\tilde{\mathbb{M}}$ of degrees of exactness of meaning, where

$$\tilde{\mathbb{M}} = \left\{ \left( m, \mu_{\tilde{\mathbb{M}}}(m) \right) \mid m \in \mho \text{ and } \mu_{\tilde{\mathbb{M}}}(m) \in [0,1] \right\} \qquad (8.3.1)$$

In other words, the meaning of any word is vague which is established by shades of meaning and processed into a fuzzy set by a membership characteristic function. The function $\mu_m(m), \forall m \in \mathbb{M} \subset \mho$ shows the degrees to which the words are exact definition of the term $\mathbb{E}$ in a language $\mathbb{L}$ .

What history seems to indicate to us in human experiences is that the best path of scientific methods is unknown, but we can claim that it exists as a potential, at least in abstract, and can be found by a continuous search. The path to knowledge is an enveloping of success-failure processes where methodological refinements proceed on approximate reasoning to help discover what works and what does not work to approach the path of best scientific methods and practices through decision-choice processes over the penumbral regions of cognition. The penumbral regions are established by the presence of inexactness, vagueness, ambiguities and information limitations. The road to the best scientific methods and practices is cloudy with vagueness, ambiguities, un-sureness, limited information and pitfalls in reasoning. It is a path of discovery that shows that it is not exact and hence the practices of science, as we have defined, are not exact and cannot follow any exact rigid methodology which is a product of thought under dynamic tensions. It is a try-and-error process where learning, error corrections and adjustments take place through the use of defective information structure as defined by incompleteness and vagueness, and supported by accepted decision-choice rationality constructed on the basis of the principles of *what works* and *what does not work*.

The path to knowledge discovery and construction is a path of learning-by-doing where the knowledge at each stage becomes a guiding light for further exploration and discovery. Knowledge discoveries and productions take place through critical thinking that is sometimes amplified by luck but not exactness. This statement does not support the anarchists position that there are no cannons of scientific methods that have not been violated at some time in history, that science grows by incorporating new theories is a myth that scientific progress occurs because scientists have no philosophy of science, and that the only principle that does not inhibit the progress of science is that *anything goes*. The only principle that facilitates progress in the knowledge production is a rationality constructed on the basis of success-failure process that has been built in a self-correction module.

As we have pointed out, the successful decision-choice processes among knowledge seekers are systematized into a framework that will provide some guidance and help to increase the chances of successful knowledge discovery. Such a framework is usually agreed upon by the social majority among knowledge seekers and specified as the framework of *scientific methods and techniques* which is contrary to the position of the anarchists. To command a general acceptance and social respect, the framework of decision-choice rationality in science is claimed to be exact in representation and reasoning. In other words, the framework acquires an ideologically protective covering and then placed into social confidence. The fact remains that there is nothing completely exact in human cognitive process. What we have is a distribution of degrees of exactness in cognition and decision-choice acceptance. The framework of scientific methods and techniques is a human construct that has its flaws, its weaknesses, its limitations and its strengths. It is obtained by a process and agreed to be appropriate in time that is subject to changes, as the house of knowledge is expanded with new rooms, new reinforced inter connecting epistemic cords, and as new areas of epistemic reflection on the ontological elements come into being.

For a continual progress towards perfection of the best methods and scientific practices, we should not allow the framework to degenerate into ideology with unconditional acceptance where the framework looses flexibility and becomes encapsulated in rigid ignorance under the classical laws of thought. The irony is that all areas of knowledge construction are mimicking this framework of exactness, including experimentation and mathematicalization, whether they fit a particular area of knowledge production or not. It is this mimicry that has given rise to mathematical economics, mathematical political science, mathematical psychology, mathematical biology and others. The failure of this understanding between the exact-inexact relational structures will move critical thinking into the sphere of tyranny of ideology that will restrict the growth of knowledge, shut out cognitive possibilities from the possibility space, imprison imagination within the ideological boundaries in the probability space and impair social progress in the space of the epistemic actual. This is not a criticism of exact science but a criticism of the ideology of exact science, and the failure of the proponents to admit that there is inexactness in exact science.

The claim of objectivity and exactness of methods of scientific research by those practicing science must be rejected even though these methods form essential parts of analytical foundations of the process of knowledge production and acceptance of what constitutes truth sometimes without any qualification. There is a circularity of reasoning in the process of explication of science. A particular set of methods and techniques of knowledge seeking is used to explicate science from non-science. The set of techniques and methods is partitioned into exact and inexact elements. Exact methods are used to explicate the exact science and inexact methods are used to explicate the inexact science where the divisions are protected by held beliefs and ideological caveats. There are times in history when some of the knowledge sectors belonging to exact sciences are discredited from being exact science and banished by other knowledge seekers from the rooms of the exact sciences in the house of knowledge. There are also times in history when some of the knowledge sectors belonging to inexact sciences are upgraded from being inexact science and brought by other knowledge seekers into the rooms of exact science or new rooms of exact science are created in the house of knowledge. These are the defining structure of the growth of knowledge.

We have argued that objectivity and exactness are not characteristics of human thought process and can not be characteristics of the epistemological space. They may be viewed as ideals to which cognitive decision-choice processes work toward their attainment. We work always with partial objectivity and exactness defined in a spectrum of fuzziness. Nonetheless, the claims of objectivity and exactness are important defining elements of cognitive foundation on which scientists draw their philosophical existence. It is this cognitive foundation that allows the scientists to establish a belief in immutability of *hard data* on the basis of exact science in *cognitive reality*. As such, it is part of exact scientific cognition to reject *soft data*, as important in the exact science, and classify knowledge-acquisition processes that rely on *soft data* as unscientific and sometimes as inexact science. Thus, exact science finds explication in *exact methods* and *hard data* that are devoid of subjectivity and human sentiments, and where the hard data are represented by exact symbols and processed with exact reasoning from the classical paradigm. This framework and adherence to it, is analytically and philosophically unfounded and restrictive to the progress and expansion of our knowledge structure and its enterprise.

In fact, it is the claim of exactness in symbolism and reasoning that forms the intense debate between the intuitionist and formalist schools of mathematics. It is also on the rejection of the claim of exactness that has given rise to the fuzzy paradigm with its logic and mathematics. Central to these epistemic disagreements is whether the information structure for cognitive processing over the epistemological space is exact or inexact. If the information structure is defective in the sense of vagueness and incomplete as we have argued and shown from the theory of the knowledge square, then what set of assumptions is acceptable to develop exact methods of reasoning?

In the name of the progress of the enterprise of the knowledge production, we must always be cognizant of the principle that we must not confuse *models of reality* as projected by cognition with reality as projected in the ontological space. Cognition is a model of reality, which we have referred to as an epistemic reality, the construction of which is guided by human decision-choice actions on the basis of subjective information that becomes accepted as objective information by collective decision and agreements. It is possible that *epistemic reality* may be an inexact and subjective projection of thought in the *epistemological space* where ontological reality, *what there is*, exists independently of thought in the *ontological space*. Such a thought, at any stage of the social knowledge development, is shrouded in vagueness, ambiguities, ideological foundation and cultural confines of thought, and hence subjective and made objective by collective decision-choice action that requires continual refinements. As such, we must resist the temptation of claiming the exact science to be free from them.

The history of cognition is culturally and ideologically shaped. In fact the progress of science, knowledge and their exactness are ideologically dependent. The degrees of ideological and cultural effects depend on the established institutions of tolerance in dealing with free pursuit of knowledge, truth and knowledge acceptance. What seems to be a universal principle is that exactness is to ontology and inexactness is to epistemology, and that fuzziness is a characteristic of science and the knowledge-production processes in the epistemological space composed of the spaces of the possibility, probability and epistemic actual. It is this universal principle that connects all areas of the knowledge production, in addition to its construction that constitute the central elements of the theory of the knowledge square as a meta-theory of knowledge. The essential elements are defective information structure, inexact symbolism, inexact information processing capacity and the claim of epistemic reality as knowledge by decision-choice action with epistemic conditionality. The toolbox for information processing consists of duality, continuum and subjective action and the fuzzy paradigm. These allow the cognitive agent to be integrated into the epistemic process in a particular language. The whole process involves the answering the following questions: 1) How is information obtained for epistemic processing?   2) What is the relative structure of quantity and quality of information? 3) How is information represented symbolically? 4) What is the culture of information processing?

# References

## R1. Bounded Rationality in Knowledge Systems

[R1.1]    Arthur, W.B.: Designing Economic Agents that Act Like Human Agents: A Behavioral Approach to Bounded Rationality. American Economic Review: Papers and Proceedings 81, 353–359 (1991)

[R1.2]    Dow, J.: Search Decisions with Limited Memory. Review of Economic Studies 58, 1–14 (1991)

[R1.3]    Gigerenzer, G., Selten, R.: Bounded Rationality: The Adaptive Toolbox. MIT Press, Cambridge (2001)

[R1.4]    Gigerenzer, G., Goldstein, D.G.: Reasoning the Fast and Frugal Way: Models of Bounded Rationality. Psychological Review 103, 650–669 (1996)

[R1.5]    Gigerenzer, G.: Bounded Rationality: Models of Fast and Frugal Inference. Swiss Journal of Economic Statistics 133, 201–218 (1997)

[R1.6]    Honkapohja, S.: Adaptive Learning and Bounded Rationality. European Economic Review 37, 587–594

[R1.7]    Lipman, B.: Information Processing and Bounded Rationality: A Survey. Canadian Jour. of Economics 28, 42–63 (1995)

[R1.8]    Lipman, B.: How to Decide How to Decide How...: Modeling Limited Rationality. Econometrica 59, 05–1125 (1991)

[R1.9]    March, J.G.: Bounded Rationality, Ambiguity and Engineering of Choice. The Bell Journal of Economics 9(2), 587–608 (1978)

[R1.10]   Neyman, A.: Bounded Rationality Justifies Cooperation in the FinitelyRepeated Prisons' Dilemma Game. Economic Letters 19, 227–229

[R1.11]   Radner, R.: Can Bounded Rationality Resolve the Prisoner's Dilemma? In: Mas-Colell, A., Hildenbrand, W. (eds.) Essays in Honor of Gerard Debreu. North-Holland, Amsterdam (1986)

[R1.12]   Rieskamp, J., et al.: Extending the Bounds of Rationality: Evidence and Theories of Preferential Choice. Journal of Economic Literature 44, 631–661 (2006)

[R1.13]   Rosenthal, R.: A Bounded-Rationality Approach to the Study of Noncooperative Games. International Journal of Game Theory 18, 273–292 (1989)

[R1.14]   Rubinstein, A.: Modeling Bounded Rationality. MIT Press, Cambridge (1998)

[R1.15]   Rubinstein, A.: New Directions in Economic Theory – Bounded Rationality. Revista Española de Economia 7, 3–15 (1990)

[R1.16]   Sargent, T.: Bounded Rationality in Macroeconomics. Clarendon, Oxford (1993)

[R1.17]   Simon, H.A.: Theories of Bounded Rationality. In: McGuire, C.B., et al. (eds.) Decision and Organization, pp. 161–176. North Holland, Amsterdam (1972)

[R1.18]    Simon, H.A.: Models of Bounded Rationality, vol. 2. MIT Press, Cambridge (1982)
[R1.19]    Simon, H.A.: From Substantive to Procedural Rationality. In: Latis, S.J. (ed.) Methods and Apprasal in Economics, pp. 129–148. Cambridge University Press, New York (1976)
[R1.20]    Starbuck, W.H.: Levels of Aspiration. Psychological Review 70, 51–60 (1963)
[R1.21]    Zemel, E.: Small Talk and Cooperation: A Note on Bounded Rationality. Journal of Economic Theory 49, 1–9
[R1.22]    Stigum, B.P., et al.: Foundation of Utility and Risk Theory with Application. D. Reidel Pub., Boston (1983)

# R 2.  Category Theory in Mathematics, Logic and Sciences

[R2.1]     Awodey, S.: Structure in Mathematics and Logic: A Categorical Perspective. Philosophia Mathematica 3, 209–237 (1996)
[R2.2]     Bell, J.L.: Category Theory and the Foundations of Mathematics. British Journal of Science 32, 349–358 (1981)
[R2.3]     Bell, J.L.: Categories, Toposes and Sets. Synthese 51, 337–393 (1982)
[R2.4]     Black, M.: The Nature of Mathematics. Littlefield, Adams and Co., Totowa (1965)
[R2.5]     Blass, A.: The Interaction Between Category and Set Theory. Mathematical Applications of Category Theory 30, 5–29 (1984)
[R2.6]     Brown, B., Woods, J. (eds.): Logical Consequence; Rival Approaches and New Studies in exact. Philosophy: Logic, Mathematics and Science, vol. II. Hermes, Oxford (2000)
[R2.7]     Butts, R.: Logic, Foundations of Mathematics and Computability. Reidel, Boston (1977)
[R2.8]     Domany, J.L., et al.: Models of Neural Networks III. Springer, New York (1996)
[R2.9]     Dompere, K.K.: Fuzzy Rationality: Methodological Critique and Unity of Classical, Bounded and Other Rationalities. Springer, New York (2009)
[R2.10]    Dompere, K.K.: The Theory of the Knowledge Square: The Fuzzy Analytical Foundations of Knowing, A Working Monograph on Philosophy of Science I. Howard University, Washington, DC (2011)
[R2.11]    Dompere, K.K.: Fuzzy Rational Foundations of Exact and Inexact Sciences, A Working Monograph on Philosophy of Science II. Howard University, Washington, DC (2011)
[R2.12]    Feferman, S.: Categorical Foundations and Foundations of Category Theory. In: Butts, R. (ed.) Logic, Foundations of Mathematics and Computability, pp. 149–169. Reidel, Boston (1977)
[R2.13]    Gray, J.W. (ed.): Mathematical Applications of Category Theory (American Mathematical Society Meeting 89th Denver Colo. 1983). American Mathematical Society, Providence (1984)
[R.2.14]   Johansson, I.: Ontological Investigations: An Inquiry into the Categories of Nature, Man, and Society. Routledge, New York (1989)
[R2.15]    Kamps, K.H., Pumplun, D., Tholen, W. (eds.): Category Theory: Proceedings of the International Conference, Gummersbach, July 6-10. Springer, New York (1982)

[R2.16]   Kent, B., Peirce, C.S.: Logic and the Classification of the Sciences. McGill-Queen's University Press, Kingston (1987)

[R2.17]   Kosko, B.: Neural Networks and Fuzzy Systems, Englewood Cliffs, NJ (1991)

[R2.18]   Landry, E.: Category Theory: the Language of Mathematics. Philosophy of Science 66 (suppl.), S14–S27

[R2.19]   Landry, E., Marquis, J.P.: Categories in Context: Historical, Foundational and Philosophical. Philosophia Mathematica 13, 1–43 (2005)

[R2.20]   Marquis, J.-P.: Three Kinds of Universals in Mathematics. In: Brown, B., Woods, J. (eds.) Logical Consequence; Rival Approaches and New Studies in Exact Philosophy. Logic, Mathematics and Science, vol. II, pp. 191–212. Hermes, Oxford (2000)

[R2.21]   McLarty, C.: Category Theory in Real Time. Philosophia Mathematica 2, 36–44 (1994)

[R2.22]   McLarty, C.: Learning from Questions on Categorical Foundations. Philosophia Mathematica 13, 44–60 (2005)

[R2.23]   Rodabaugh, S., et al. (eds.): Application of Category Theory to Fuzzy Subsets. Kluwer, Boston (1992)

[R2.24]   Taylor, J.G. (ed.): Mathematical Approaches to Neural Networks. North-Holland, New York (1993)

[R2.25]   Van Benthem, J., et al. (eds.): The Age of Alternative Logics: Assessing Philosophy of Logic and Mathematics Today. Springer, New York (2006)

## R3. Fuzzy Logic in Knowledge Production

[R3.1]    Baldwin, J.F.: A New Approach to Approximate Reasoning Using a Fuzzy Logic. Fuzzy Sets and Systems 2(4), 309–325 (1979)

[R3.2]    Baldwin, J.F.: Fuzzy Logic and Fuzzy Reasoning. Intern. J. Man-Machine Stud. 11, 465–480 (1979)

[R3.3]    Baldwin, J.F.: Fuzzy Logic and Its Application to Fuzzy Reasoning. In: Gupta, M.M., et al. (eds.) Advances in Fuzzy Set Theory and Applications, pp. 96–115. North-Holland, New York (1979)

[R3.4]    Baldwin, J.F., et al.: Fuzzy Relational Inference Language. Fuzzy Sets and Systems 14(2), 155–174 (1984)

[R3.5]    Baldsin, J., Pilsworth, B.W.: Axiomatic Approach to Implication For Approximate Reasoning With Fuzzy Logic. Fuzzy Sets and Systems 3(2), 193–219 (1980)

[R3.6]    Baldwin, J.F., et al.: The Resolution of Two Paradoxes by Approximate Reasoning Using A Fuzzy Logic. Synthese 44, 397–420 (1980)

[R3.7]    Fukami, S., et al.: Some Considerations On Fuzzy Conditional Inference. Fuzzy Sets and Systems 4(3), 243–273 (1980)

[R3.8]    Gaines, B.R.: Fuzzy Reasoning and the Logic of Uncertainty. In: Proc. 6th International Symp. of Multiple-Valued Logic, pp. 179–188. IEEE 76CH 1111-4C (1976)

[R3.9]    Gaines, B.R.: Foundations of Fuzzy Reasoning. Inter. Jour. of Man-Machine Studies 8, 623–668 (1976)

[R3.10]   Gaines, B.R.: Foundations of Fuzzy Reasoning. In: Gupta, M.M., et al. (eds.) Fuzzy Information and Decision Processes, pp. 19–75. North-Holland, New York (1982)

[R3.11]   Gaines, B.R.: Precise Past, Fuzzy Future. International Journal of Man-Machine Stududies 19(1), 117–134 (1983)

[R3.12]   Gains, B.R.: Fuzzy and Probability Uncertainty Logics. Information and Control 38, 154–169 (1978)

[R3.13]   Gains, B.R.: Modeling Practical Reasoning. Intern. Jour. of Intelligent Systems 8(1), 51–70 (1993)

[R3.14]   Gaines, B.R.: Łukasiewicz Logic and Fuzzy Set Theory. International Jour. of Man-Machine Studies 8, 313–327 (1976)

[R3.15]   Giles, R.: Lukasiewics Logic and Fuzzy Set Theory. Intern. J. Man-Machine Stud. 8, 313–327 (1976)

[R3.16]   Giles, R.: Formal System for Fuzzy Reasoning. Fuzzy Sets and Systems 2(3), 233–257 (1979)

[R3.17]   Ginsberg, M.L. (ed.): Readings in Non-monotonic Reason. Morgan Kaufmann, Los Altos (1987)

[R3.18]   Goguen, J.A.: The Logic of Inexact Concepts. Synthese 19, 325–373 (1969)

[R3.19]   Gottinger, H.W.: Towards a Fuzzy Reasoning in the Behavioral Science. Cybernetica 16(2), 113–135 (1973)

[R3.20]   Gottinger, H.W.: Some Basic Issues Connected With Fuzzy Analysis. In: Klaczro, H., Muller, N. (eds.) Systems Theory in Social Sciences, pp. 323–325. Birkhauser Verlag, Basel (1976)

[R3.21]   Gottwald, S.: Fuzzy Propositional Logics. Fuzzy Sets and Systems 3(2), 181–192 (1980)

[R3.22]   Gupta, M.M., et al.: Approximate Reasoning In Decision Analysis. North Holland, New York (1982)

[R3.23]   Ulrich, H., Klement, E.P.: Non-Clasical Logics and their Applications to Fuzzy Subsets: A Handbook of the Mathematical Foundations of Fuzzy Set Theory. Kluwer, Boston (1995)

[R3.24]   Kaipov, V.K., et al.: Classification in Fuzzy Environments. In: Gupta, M.M., et al. (eds.) Advances in Fuzzy Set Theory and Applications, pp. 119–124. North-Holland, New York (1979)

[R3.25]   Kaufman, A.: Progress in Modeling of Human Reasoning of Fuzzy Logic. In: Gupta, M.M., et al. (eds.) Fuzzy Information and Decision Process, pp. 11–17. North-Holland, New York (1982)

[R3.26]   Lakoff, G.: Hedges: A Study in Meaning Criteria and the Logic of Fuzzy Concepts. Jour. Philos. Logic 2, 458–508 (1973)

[R3.27]   Lee, E.T., et al.: Some Properties of Fuzzy Logic. Information and Control 19, 417–431 (1971)

[R3.28]   Lee, R.C.T.: Fuzzy Logic and the Resolution Principle. Jour. of Assoc. Comput. Mach. 19, 109–119 (1972)

[R3.29]   LeFaivre, R.A.: The Representation of Fuzzy Knowledge. Jour. of Cybernetics 4, 57–66 (1974)

[R3.30]   Mitra, S., Pal, S.K.: Logical Operation based Fuzzy MLP for Classification and Rule Generation. Neural Networks 7(2), 353–373 (1994)

[R3.31]   Mizumoto, M.: Fuzzy Conditional Inference Under Max- ⊙ Composition. Information Sciences 27(3), 183–207 (1982)

[R3.32]   Mizumoto, M., et al.: Several Methods For Fuzzy Conditional Inference. In: Proc. of IEEE Conf. on Decision and Control, Florida, December 12-14, pp. 777–782 (1979)

[R3.33]  Montero, F.J.: Measuring the Rationality of a Fuzzy Preference Relation. Busefal. 26, 75–83 (1986)

[R3.34]  Morgan, C.G.: Methods for Automated Theorem Proving in Non- Classical Logics. IEEE Trans. Compt. C-25, 852–862 (1976)

[R3.35]  Negoita, C.V.: Representation Theorems for Fuzzy Concepts. Kybernetes 4, 169–174 (1975)

[R3.36]  Nguyen, H.T., Walker, E.A.: A First Course In Fuzzy Logic. CRC Press, Boca Raton (1997)

[R3.37]  Nowakowska, M.: Methodological Problems of Measurements of Fuzzy Concepts in Social Sciences. Behavioral Sciences 22(2), 107–115 (1977)

[R3.38]  Pinkava, V.: Fuzzification of Binary and Finite Multivalued Logical Calculi. Intern. Jour. Man-Machine Stud. 8, 171–730 (1976)

[R3.39]  Skala, H.J.: Non-Archimedean Utility Theory. D. Reidel, Dordrecht (1975)

[R3.40]  Skala, H.J.: On Many-Valued Logics, Fuzzy Sets, Fuzzy Logics and Their Applications. Fuzzy Sets and Systems 1(2), 129–149 (1978)

[R3.41]  Sugeno, M., Takagi, T.: Multi-Dimensional Fuzzy Reasoning. Fuzzy Sets and Systems 9(3), 313–325 (1983)

[R3.42]  Tan, S.K., et al.: Fuzzy Inference Relation Based on the Theory of Falling Shadows. Fuzzy Sets and Systems 53(2), 179–188 (1993)

[R3.43]  Thornber, K.K.: A Key to Fuzzy Logic Inference. Intern. Jour. of Approximate Reasoning 8(2), 105–129 (1993)

[R3.44]  Tong, R.M., et al.: A Critical Assessment of Truth Functional Modification and its Use in Approximate Reasoning. Fuzzy Sets and Systems 7(1), 103–108 (1982)

[R3.45]  Van Fraassen, B.C.: Comments: Lakoff's Fuzzy Propositional Logic. In: Hockney, D., et al. (eds.) Contemporary Research in Philosophical Logic and Linguistic Semantics Holland, Reild, pp. 273–277 (1975)

[R3.46]  Whalen, T., et al.: Usuality, Regularity, and Fuzzy Set Logic. Intern. Jour. of Approximate Reasoning 6(4), 481–504 (1992)

[R3.47]  Yager, R.R., et al. (eds.): An Introduction to Fuzzy Logic Applications in Intelligent Systems. Kluwer, Boston (1992)

[R3.48]  Yan, J., et al.: Using Fuzzy Logic: Towards Intelligent Systems. Prentice-Hall, Englewood Cliffs (1994)

[R3.49]  Ying, M.S.: Some Notes on Multidimensional Fuzzy Reasoning. Cybernetics and Systems 19(4), 281–293 (1988)

[R3.50]  Zadeh, L.A.: Quantitative Fuzzy Semantics. Inform. Science 3, 159–176 (1971)

[R3.51]  Zadeh, L.A.: A Fuzzy Set Interpretation of Linguistic Hedges. Jour. Cybernetics 2, 4–34 (1972)

[R3.52]  Zadeh, L.A.: Fuzzy Logic and Its Application to Approximate Reasoning. In: Information Processing 1974, Proc. IFIP Congress, vol. 74(3), pp. 591–594. North Holland, New York (1974)

[R3.53]  Zadeh, L.A.: The Concept of a Linguistic Variable and Its Application to Approximate Reasoning. In: Fu, K.S., et al. (eds.) Learning Systems and Intelligent Robots, pp. 1–10. Plenum Press, New York (1974)

[R3.54]  Zadeh, L.A.: Fuzzy Logic and Approximate Reasoning. Syntheses 30, 407–428 (1975)

[R3.55]  Zadeh, L.A., et al. (eds.): Fuzzy Logic for the Management of Uncertainty. Wily and Sons, New York (1992)

[R3.56]   Zadeh, L.A., et al. (eds.): Fuzzy Sets and Their Applications to Cognitive and
          Decision Processes. Academic Press, New York (1974)
[R3.57]   Zadeh, L.A.: The Birth and Evolution of Fuzzy Logic. Intern. Jour. of General
          Systems 17(2-3), 95–105 (1990)

## R4.  Fuzzy Mathematics in Approximate Reasoning
##       under Conditions of Inexactness and Vagueness

[R4.1]    Bandler, W., et al.: Fuzzy Power Sets and Fuzzy Implication Operators. Fuzzy
          Sets and Systems 4(1), 13–30 (1980)
[R4.2]    Banon, G.: Distinction between Several Subsets of Fuzzy Measures. Fuzzy Sets
          and Systems 5(3), 291–305 (1981)
[R4.3]    Bellman, R.E.: Mathematics and Human Sciences. In: Wilkinson, J., et al. (eds.)
          The Dynamic Programming of Human Systems, pp. 11–18. MSS Information
          Corp., New York (1973)
[R4.4]    Bellman, R.E., Glertz, M.: On the Analytic Formalism of the Theory of Fuzzy
          Sets. Information Science 5, 149–156 (1973)
[R4.5]    Buckley, J.J.: The Fuzzy Mathematics of Finance. Fuzzy Sets and
          Systems 21(3), 257–273 (1987)
[R4.6]    Butnariu, D.: Fixed Points For Fuzzy Mapping. Fuzzy Sets and Systems 7(2),
          191–207 (1982)
[R4.7]    Butnariu, D.: Decompositions and Range For Additive Fuzzy Measures. Fuzzy
          Sets and Systems 10(2), 135–155 (1983)
[R4.8]    Cerruti, U.: Graphs and Fuzzy Graphs. In: Fuzzy Information and Decision
          Processes, pp. 123–131. North-Holland, New York (1982)
[R4.9]    Chakraborty, M.K., et al.: Studies in Fuzzy Relations Over Fuzzy Subsets.
          Fuzzy Sets and Systems 9(1), 79–89 (1983)
[R4.10]   Chang, C.L.: Fuzzy Topological Spaces. J. Math. Anal. and Applications 24,
          182–190 (1968)
[R4.11]   Chang, S.S.L.: Fuzzy Mathematics, Man and His Environment. IEEE
          Transactions on Systems, Man and Cybernetics, SMC-2, 92–93 (1972)
[R4.12]   Chang, S.S.L., et al.: On Fuzzy Mathematics and Control. IEEE Transactions,
          System, Man and Cybernetics, SMC-2, 30–34 (1972)
[R4.13]   Chang, S.S.: Fixed Point Theorems for Fuzzy Mappings. Fuzzy Sets and
          Systems 17, 181–187 (1985)
[R4.14]   Chapin, E.W.: An Axiomatization of the Set Theory of Zadeh. Notices,
          American Math. Society, 687-02-4 754 (1971)
[R4.15]   Chaudhury, A.K., Das, P.: Some Results on Fuzzy Topology on Fuzzy Sets.
          Fuzzy Sets and Systems 56, 331–336 (1993)
[R4.16]   Cheng-Zhong, L.: Generalized Inverses of Fuzzy Matrix. In: Gupta, M.M., et al.
          (eds.) Approximate Reasoning In Decision Analysis, pp. 57–60. North Holland,
          New York (1982)
[R4.17]   Chitra, H., Subrahmanyam, P.V.: Fuzzy Sets and Fixed Points. Jour. of
          Mathematical Analysis and Application 124, 584–590 (1987)
[R4.18]   Cohn, D.L.: Measure Theory. Birkhauser, Boston (1980)
[R4.19]   Cohen, P.J., Hirsch, R.: Non-Cantorian Set Theory. Scientific America, 101–116
          (December 1967)

[R4.20]   Czogala, J., et al.: Fuzzy Relation Equations On a Finite Set. Fuzzy Sets and Systems 7(1), 89–101 (1982)

[R4.21]   Das, P.: Fuzzy Topology on Fuzzy Sets: Product Fuzzy Topology and Fuzzy Topological Groups. Fuzzy Sets and Systems 100, 367–372 (1998)

[R4.22]   Dinola, A., et al.: The Mathematics of Fuzzy Systems. Verlag TUV Rheinland, Koln (1986)

[R4.23]   Dombi, J.: A General Class of Fuzzy Operators, the DeMorgan Class of Fuzzy Operators and Fuzzy Measures Induced by Fuzzy Operators. Fuzzy Sets and Systems 8(2), 149–163 (1982)

[R4.24]   Dubois, D., Prade, H.: Fuzzy Sets and Systems. Academic Press, New York (1980)

[R4.25]   Dubois: Fuzzy Real Algebra: Some Results. Fuzzy Sets and Systems 2(4), 327–348 (1979)

[R4.26]   Dubois, D., Prade, H.: Gradual Inference Rules in Approximate Reasoning. Information Sciences 61(1-2), 103–122 (1992)

[R4.27]   Dubois, D., Prade, H.: On the Combination of Evidence in various Mathematical Frameworks. In: Flamm, J., Luisi, T. (eds.) Reliability Data Collection and Analysis, pp. 213–241. Kluwer, Boston (1992)

[R4.28]   Dubois, D., Prade, H.: Fuzzy Sets and Probability: Misunderstanding, Bridges and Gaps. In: Proc. Second IEEE Intern. Conf. on Fuzzy Systems, San Francisco, pp. 1059–1068 (1993)

[R4.29]   Dubois, D., Prade, H.: A Survey of Belief Revision and Updating Rules in Various Uncertainty Models. Intern. J. of Intelligent Systems 9(1), 61–100 (1994)

[R4.30]   Erceg, M.A.: Functions, Equivalence Relations, Quotient Spaces and Subsets in Fuzzy Set Theory. Fuzzy Sets and Systems 3(1), 79–92 (1980)

[R4.31]   Feng, Y.-J.: A Method Using Fuzzy Mathematics to Solve the Vectormaximum Problem. Fuzzy Sets and Systems 9(2), 129–136 (1983)

[R4.32]   Filev, D.P., et al.: A Generalized Defuzzification Method via Bag Distributions. Intern. Jour. of Intelligent Systems 6(7), 687–697 (1991)

[R4.33]   Foster, D.H.: Fuzzy Topological Groups. Journal of Math. Analysis and Applications 67, 549–564 (1979)

[R4.34]   Goetschel Jr., R., et al.: Topological Properties of Fuzzy Number. Fuzzy Sets and Systems 10(1), 87–99 (1983)

[R4.35]   Goodman, I.R.: Fuzzy Sets As Random Level Sets: Implications and Extensions of the Basic Results. In: Lasker, G.E. (ed.) Applied Systems and Cybernetics, Fuzzy Sets and Systems, vol. VI, pp. 2756–2766. Pergamon Press, New York (1981)

[R4.36]   Goodman, I.R.: Fuzzy Sets As Equivalence Classes of Random Sets. In: Yager, R.R. (ed.) Fuzzy Set and Possibility Theory: Recent Development, pp. 327–343. Pergamon Press, New York (1992)

[R4.37]   Gupta, M.M., et al. (eds.): Fuzzy Antomata and Decision Processes. North-Holland, New York (1977)

[R4.38]   Gupta, M.M., Sanchez, E. (eds.): Fuzzy Information and Decision Processes. North-Holland, New York (1982)

[R4.39]   Higashi, M., Klir, G.J.: On measure of fuzziness and fuzzy complements. Intern. J. of General Systems 8(3), 169–180 (1982)

[R4.40]   Higashi, M., Klir, G.J.: Measures of uncertainty and information based on possibility distributions. International Journal of General Systems 9(1), 43–58 (1983)

[R4.41]   Higashi, M., Klir, G.J.: On the notion of distance representing information closeness: Possibility and probability distributions. Intern. J. of General Systems 9(2), 103–115 (1983)

[R4.42]   Higashi, M., Klir, G.J.: Resolution of finite fuzzy relation equations. Fuzzy Sets and Systems 13(1), 65–82 (1984)

[R4.43]   Higashi, M., Klir, G.J.: Identification of fuzzy relation systems. IEEE Trans. on Systems, Man, and Cybernetics 14(2), 349–355 (1984)

[R4.44]   Ulrich, H.: A Mathematical Theory of Uncertainty. In: Yager, R.R. (ed.) Fuzzy Set and Possibility Theory: Recent Developments, pp. 344–355. Pergamon, New York (1982)

[R4.45]   Jin-Wen, Z.: A Unified Treatment of Fuzzy Set Theory and Boolean Valued Set theory: Fuzzy Set Structures and Normal Fuzzy Set Structures. Jour. Math. Anal. and Applications 76(1), 197–301 (1980)

[R4.46]   Kandel, A.: Fuzzy Mathematical Techniques with Applications. Addison-Wesley, Reading (1986)

[R4.47]   Kandel, A., Byatt, W.J.: Fuzzy Processes. Fuzzy Sets and Systems 4(2), 117–152 (1980)

[R4.48]   Kaufmann, A., Gupta, M.M.: Introduction to fuzzy arithmetic: Theory and applications. Van Nostrand, New York (1991)

[R4.49]   Kaufmann, A.: Introduction to the Theory of Fuzzy Subsets, vol. 1. Academic Press, New York (1975)

[R4.50]   Kaufmann, A.: Theory of Fuzzy Sets. Merson Press, Paris (1972)

[R4.51]   Kaufmann, A., et al.: Fuzzy Mathematical Models in Engineering and Management Science. North-Holland, New York (1988)

[R4.52]   Kim, K.H., et al.: Generalized Fuzzy Matrices. Fuzzy Sets and Systems 4(3), 293–315 (1980)

[R4.53]   Klement, E.P.: Fuzzy $\sigma$-Algebras and Fuzzy Measurable Functions. Fuzzy Sets and Systems 4, 83–93 (1980)

[R4.54]   Klement, E.P.: Characterization of Finite Fuzzy Measures Using Markoff-kernels. Journal of Math. Analysis and Applications 75, 330–339 (1980)

[R4.55]   Klement, E.P.: Construction of Fuzzy $\sigma$-Algebras Using Triangular Norms. Journal of Math. Analysis and Applications 85, 543–565 (1982)

[R4.56]   Klement, E.P., Schwyhla, W.: Correspondence Between Fuzzy Measures and Classical Measures. Fuzzy Sets and Systems 7(1), 57–70 (1982)

[R4.57]   Klir, G., Bo, Y.: Fuzzy Sets and Fuzzy Logic. Prentice-Hall, Upper Saddle River (1995)

[R4.58]   Kokawa, M., et al.: Fuzzy-Theoretical Dimensionality Reduction Method of Multi-Dimensional Quality. In: Gupta, M.M., Sanchez, E. (eds.) Fuzzy Information and Decision Processes, pp. 235–250. North-Holland, New York (1982)

[R4.59]   Kramosil, I., et al.: Fuzzy Metrics and Statistical Metric Spaces. Kybernetika 11, 336–344 (1975)

[R4.60]   Kruse, R.: On the Construction of Fuzzy Measures. Fuzzy Sets and Systems 8(3), 323–327 (1982)

[R4.61]   Kruse, R., et al.: Foundations of Fuzzy Systems. John Wiley and Sons, New York (1994)

[R4.62] Lasker, G.E. (ed.): Applied Systems and Cybernetics, vol. VI. Pergamon Press, New York (1981)

[R4.63] Lientz, B.P.: On Time Dependent Fuzzy Sets. Inform. Science 4, 367–376 (1972)

[R4.64] Lowen, R.: Fuzzy Uniform Spaces. Jour. Math. Anal. Appl. 82(2), 367–376 (1981)

[R4.65] Lowen, R.: On the Existence of Natural Non-Topological Fuzzy Topological Space. Haldermann Verlag, Berlin (1986)

[R4.66] Martin, H.W.: Weakly Induced Fuzzy Topological Spaces. Jour. Math. Anal. and Application 78, 634–639 (1980)

[R4.67] Michalek, J.: Fuzzy Topologies. Kybernetika 11, 345–354 (1975)

[R4.68] Mizumoto, M., Tanaka, K.: Some Properties of Fuzzy Numbers. In: Gupta, M.M., et al. (eds.) Advances in Fuzzy Sets Theory and Applications, pp. 153–164. North Holland, Amsterdam (1979)

[R4.69] Negoita, C.V., et al.: Applications of Fuzzy Sets to Systems Analysis. Wiley and Sons, New York (1975)

[R4.70] Negoita, C.V.: Representation Theorems for Fuzzy Concepts. Kybernetes 4, 169–174 (1975)

[R4.71] Negoita, C.V., et al.: On the State Equation of Fuzzy Systems. Kybernetes 4, 213–216 (1975)

[R4.72] Negoita, C.V.: Fuzzy Sets in Topoi. Fuzzy Sets and Systems 8(1), 93–99 (1982)

[R4.73] Netto, A.B.: Fuzzy Classes. Notices, American Mathematical Society 68T-H28, 945 (1968)

[R4.74] Nguyen, H.T.: Possibility Measures and Related Topics. In: Gupta, M.M., et al. (eds.) Approximate Reasoning in Decision Analysis, pp. 197–202. North Holland, New York (1982)

[R4.75] Nowakowska, M.: Some Problems in the Foundations of Fuzzy Set Theory. In: Gupta, M.M., et al. (eds.) Approximate Reasoning in Decision Analysis, pp. 349–360. North Holland, New York (1982)

[R4.76] Ovchinnikov, S.V.: Structure of Fuzzy Binary Relations. Fuzzy Sets and Systems 6(2), 169–195 (1981)

[R4.77] Pedrycz, W.: Fuzzy Relational Equations with Generalized Connectives and Their Applications. Fuzzy Sets and Systems 10(2), 185–201 (1983)

[R4.78] Raha, S., et al.: Analogy Between Approximate Reasoning and the Method of Interpolation. Fuzzy Sets and Systems 51(3), 259–266 (1992)

[R4.79] Ralescu, D.: Toward a General Theory of Fuzzy Variables. Jour. of Math. Analysis and Applications 86(1), 176–193 (1982)

[R4.80] Rao, M.B., et al.: Some Comments On Fuzzy Variables. Fuzzy Sets and Systems 6(2), 285–292 (1981)

[R4.81] Rodabaugh, S.E.: Fuzzy Arithmetic and Fuzzy Topology. In: Lasker, G.E. (ed.) Applied Systems and Cybernetics, Fuzzy Sets and Systems, vol. VI, pp. 2803–2807. Pergamon Press, New York (1981)

[R4.82] Rodabaugh, S., et al. (eds.): Application of Category Theory to Fuzzy Subsets. Kluwer, Boston (1992)

[R4.83] Roubens, M., et al.: Linear Fuzzy Graphs. Fuzzy Sets and Systems 10(1), 798–806 (1983)

[R4.84] Rosenfeld, A.: Fuzzy Groups. Jour. Math. Anal. Appln. 35, 512–517 (1971)

[R4.85]    Rosenfeld, A.: Fuzzy Graphs. In: Zadeh, L.A., et al. (eds.) Fuzzy Sets and Their Applications to Cognitive and Decision Processes, pp. 77–95. Academic Press, New York (1974)

[R4.86]    Ruspini, E.H.: Recent Developments in Mathematical Classification Using Fuzzy Sets. In: Lasker, G.E. (ed.) Applied Systems and Cybernetics, Fuzzy Sets and Systems, vol. VI, pp. 2785–2790. Pergamon Press, New York (1981)

[R4.87]    Santos, E.S.: Maximin, Minimax and Composite Sequential Machines. Jour. Math. Anal. and Appln. 24, 246–259 (1968)

[R4.88]    Santos, E.S.: Fuzzy Algorithms. Inform. and Control 17, 326–339 (1970)

[R4.89]    Sarkar, M.: On Fuzzy Topological Spaces. Jour. Math. Anal. Appln. 79, 384–394 (1981)

[R4.90]    Slowinski, R., Teghem, J.: Stochastic versus Fuzzy Approaches to Multiobjective Mathematical Programming Under Uncertainty. Kluwer, Dordrecht (1990)

[R4.91]    Stein, N.E., Talaki, K.: Convex Fuzzy Random Variables. Fuzzy Sets and Systems 6(3), 271–284 (1981)

[R4.92]    Sugeno, M.: Inverse Operation of Fuzzy Integrals and Conditional Fuzzy Measures. Transactions SICE 11, 709–714 (1975)

[R4.93]    Yager, R.R., Filver, D.P.: Essentials of Fuzzy Modeling and Control. John Wiley and Sons, New York (1994)

[R4.94]    Triantaphyllon, E., et al.: The Problem of Determining Membership Values in Fuzzy Sets in Real World Situations. In: Brown, D.E., et al. (eds.) Operations Research and Artificial Intelligence: The Integration of Problem-solving Strategies, pp. 197–214. Kluwer, Boston (1990)

[R4.95]    Tsichritzis, D.: Participation Measures. Jour. Math. Anal. and Appln. 36, 60–72 (1971)

[R4.96]    Tsichritzis, D.: Approximation and Complexity of Functions on the Integers. Inform. Science 4, 70–86 (1971)

[R4.97]    Turksens, I.B.: Four Methods of Approximate Reasoning with Interval-Valued Fuzzy Sets. Intern. Journ. of Approximate Reasoning 3(2), 121–142 (1989)

[R4.98]    Turksen, I.B.: Measurement of Membership Functions and Their Acquisition. Fuzzy Sets and Systems 40(1), 5–38 (1991)

[R4.99]    Wang, L.X.: Adaptive Fuzzy Sets and Control: Design and Stability Analysis. Prentice-Hall, Englewood Cliffs (1994)

[R4.100]   Wang, P.P. (ed.): Advances in Fuzzy Sets, Possibility Theory, and Applications. Plenum Press, New York (1983)

[R4.101]   Wang, P.P. (ed.): Advances in Fuzzy Theory and Technology, vol. 1. Bookwright Press, Durham (1992)

[R4.102]   Wang, Z., Klir, G.: Fuzzy Measure Theory. Plenum Press, New York (1992)

[R4.103]   Wang, P.Z., et al. (eds.): Between Mind and Computer: Fuzzy Science and Engineering. World Scientific Press, Singapore (1993)

[R4.104]   Wang, P.Z.: Contactability and Fuzzy Variables. Fuzzy Sets and Systems 8(1), 81–92 (1982)

[R4.105]   Wang, S.: Generating Fuzzy Membership Functions: A Monotonic Neural Network Model. Fuzzy Sets and Systems 61(1), 71–82 (1994)

[R4.106]   Wierzchon, S.T.: An Algorithm for Identification of Fuzzy Measure. Fuzzy Sets and Systems 9(1), 69–78 (1983)

[R4.107]   Wong, C.K.: Fuzzy Topology: Product and Quotient Theorems. Journal of Math. Analysis and Applications 45, 512–521 (1974)

[R4.108]  Wong, C.K.: Fuzzy Points and Local Properties of Fuzzy Topology. Jour. Math. Anal. and Appln. 46, 316–328 (1987)

[R4.109]  Wong, C.K.: Categories of Fuzzy Sets and Fuzzy Topological Spaces. Jour. Math. Anal. and Appln. 53, 704–714 (1976)

[R4.110]  Yager, R.R.: On the Lack of Inverses in Fuzzy Arithmetic. Fuzzy Sets and Systems 4(1), 73–82 (1980)

[R4.111]  Yager, R.R. (ed.): Fuzzy Set and Possibility Theory: Recent Development. Pergamon Press, New York (1992)

[R4.112]  Yager, R.R.: Fuzzy Subsets with Uncertain Membership Grades. IEEE Transactions on Systems, Man and Cybernetics 14(2), 271–275 (1984)

[R4.113]  Yager, R.R., et al. (eds.): Fuzzy Sets, Neural Networks, and Soft Computing. Nostrand Reinhold, New York (1994)

[R4.114]  Yager, R.R.: On the Theory of Fuzzy Bags. Intern. Jour. of General Systems 13(1), 23–37 (1986)

[R4.115]  Yager, R.R.: Cardinality of Fuzzy Sets via Bags. Mathematical Modelling 9(6), 441–446 (1987)

[R4.116]  Zadeh, L.A.: A Computational Theory of Decompositions. Intern. Jour. of Intelligent Systems 2(1), 39–63 (1987)

[R4.117]  Zadeh, L.A., et al.: Fuzzy Logic for the Management of Uncertainty. John Wiley, New York (1992)

[R4.118]  Zimmerman, H.J.: Fuzzy Set Theory and Its Applications. Kluwer, Boston (1985)

## R5. Fuzzy Optimization, Decision-Choices and Approximate Reasoning in Sciences

[R5.1]  Bose, R.K., Sahani, D.: Fuzzy Mappings and Fixed Point Theorems. Fuzzy Sets and Systems 21, 53–58 (1987)

[R5.2]  Buckley, J.J.: Fuzzy Programming And the Pareto Optimal Set. Fuzzy Set and Systems 10(1), 57–63 (1983)

[R5.3]  Butnariu, D.: Fixed Points for Fuzzy Mappings. Fuzzy Sets and Systems 7, 191–207 (1982)

[R5.4]  Carlsson, G.: Solving Ill-Structured Problems Through Well Structured Fuzzy Programming. In: Brans, J.P. (ed.) Operation Research, vol. 81, pp. 467–477. North-Holland, Amsterdam (1981)

[R5.5]  Carlsson, C.: Tackling an AMDM – Problems with the Help of Some Results From Fuzzy Set Theory. European Journal of Operational Research 10(3), 270–281 (1982)

[R5.6]  Cerny, M.: Fuzzy Approach to Vector Optimization. Intern. Jour. of General Systems 20(1), 23–29

[R5.7]  Chang, S.S.: Fixed Point Theorems for Fuzzy Mappings. Fuzzy Sets and Systems 17, 181–187 (1985)

[R5.8]  Chang, S.Y., et al.: Modeling To Generate Alternatives: A Fuzzy Approach. Fuzzy Sets and Systems 9(2), 137–151 (1983)

[R5.9]  Dubois, D.: An Application of Fuzzy Arithmetic to the Optimization of Industrial Machining Processes. Mathematical Modelling 9(6), 461–475 (1987)

[R5.10]   Edwards, W.: The Theory of Decision Making. Psychological Bulletin 51, 380–417 (1954)

[R5.11]   Eaves, B.C.: Computing Kakutani Fixed Points. Journal of Applied Mathematics 21, 236–244 (1971)

[R5.12]   Feng, Y.J.: A Method Using Fuzzy Mathematics to Solve the Vector Maxim Problem. Fuzzy Set and Systems 9(2), 129–136 (1983)

[R5.13]   Hannan, E.L.: Linear Programming with Multiple Fuzzy Goals. Fuzzy Sets and Systems 6(3), 235–248 (1981)

[R5.14]   Heilpern, S.: Fuzzy Mappings and Fixed Point Theorem. Journal of Mathematical Analysis and Applications 83, 566–569 (1981)

[R.15]    Ignizio, J.P., et al.: Fuzzy Multicriteria Integer Programming via Fuzzy Generalized Networks. Fuzzy Sets and Systems 10(3), 261–270 (1983)

[R5.16]   Kacprzyk, J., et al. (eds.): Optimization Models Using Fuzzy Sets and Possibility Theory. D. Reidel, Boston (1987)

[R5.17]   Kakutani, S.: A Generalization of Brouwer's Fixed Point Theorem. Duke Mathematical Journal 8, 416–427 (1941)

[R5.18]   Kaleva, O.: A Note on Fixed Points for Fuzzy Mappings. Fuzzy Sets and Systems 15, 99–100 (1985)

[R5.19]   Kandel, A.: On Minimization of Fuzzy Functions. IEEE Trans. Comp. C-22, 826–832 (1973)

[R5.20]   Kandel, A.: On the Minimization of Incompletely Specified Fuzzy Functions. Information, and Control 26, 141–153 (1974)

[R5.21]   Lai, Y., et al.: Fuzzy Mathematical Programming. Springer, New York (1992)

[R5.22]   Leberling, H.: On Finding Compromise Solution in Multcriteria Problems, Using the Fuzzy Min-Operator. Fuzzy Set and Systems 6(2), 105–118 (1981)

[R5.23]   Lee, E.S., et al.: Fuzzy Multiple Objective Programming and Compromise Programming with Pareto Optimum. Fuzzy Sets and Systems 53(3), 275–288 (1993)

[R5.24]   Lodwick, W.A., Kacprzyk, J.: Fuzzy Optimization: Recent Advances and Applications. STUDFUZZ, vol. 254. Springer, Heidelberg (2010)

[R5.25]   Lowen, R.: Convex Fuzzy Sets. Fuzzy Sets and Systems 3, 291–310 (1980)

[R5.26]   Luhandjula, M.K.: Compensatory Operators in Fuzzy Linear Programming with Multiple Objectives. Fuzzy Sets and Systems 8(3), 245–252 (1982)

[R5.27]   Luhandjula, M.K.: Linear Programming Under Randomness and Fuzziness. Fuzzy Sets and Systems 10(1), 45–54 (1983)

[R5.28]   Negoita, C.V., et al.: Fuzzy Linear Programming and Tolerances in Planning. Econ. Group Cybernetic Studies 1, 3–15 (1976)

[R5.29]   Negoita, C.V., Stefanescu, A.C.: On Fuzzy Optimization. In: Gupta, M.M., et al. (eds.) Approximate Reasoning in Decision Analysis, pp. 247–250. North-Holland, New York (1982)

[R5.30]   Negoita, C.V.: The Current Interest in Fuzzy Optimization. Fuzzy Sets and Systems 6(3), 261–270 (1981)

[R5.31]   Negoita, C.V., et al.: On Fuzzy Environment in Optimization Problems. In: Rose, J., et al. (eds.) Modern Trends in Cybernetics and Systems, pp. 13–24. Springer, Berlin (1977)

[R5.32]   Orlovsky, S.A.: On Formulation of General Fuzzy Mathematical Problem. Fuzzy Sets and Systems 3, 311–321 (1980)

[R5.33]   Ostasiewicz, W.: A New Approach to Fuzzy Programming. Fuzzy Sets and Systems 7(2), 139–152 (1982)

[R5.34]   Pollatschek, M.A.: Hieranchical Systems and Fuzzy-Set Theory. Kybernetes 6, 147–151 (1977)

[R5.35]   Ponsard, G.: Partial Spatial Equilibra With Fuzzy Constraints. Journal of Regional Science 22(2), 159–175 (1982)

[R5.36]   Prade, M.: Operations Research with Fuzzy Data. In: Want, P.P., et al. (eds.) Fuzzy Sets, pp. 155–170. Plenum, New York (1980)

[R5.37]   Ralescu, D.: Optimization in a Fuzzy Environment. In: Gupta, M.M., et al. (eds.) Advances in Fuzzy Set Theory and Applications, pp. 77–91. North-Holland, New York (1979)

[R5.38]   Ralescu, D.A.: Orderings, Preferences and Fuzzy Optimization. In: Rose, J. (ed.) Current Topics in Cybernetics and Systems. Springer, Berlin (1978)

[R5.39]   Tanaka, K., et al.: Fuzzy Programs and Their Execution. In: Zadeh, L.A., et al. (eds.) Fuzzy Sets and Their Applications to Cognitive and Decision Processes, pp. 41–76 (1974)

[R5.40]   Tanaka, H., et al.: On Fuzzy-Mathematical Programming. Journal of Cybernetics 3(4), 37–46 (1974)

[R5.41]   Vira, J.: Fuzzy Expectation Values in Multistage Optimization Problems. Fuzzy Sets and Systems 6(2), 161–168 (1981)

[R5.42]   Verdegay, J.L.: Fuzzy Mathematical Programming. In: Gupta, M.M., et al. (eds.) Fuzzy Information and Decision Processes, pp. 231–238. North-Holland, New York (1982)

[R5.43]   Warren, R.H.: Optimality in Fuzzy Topological Polysystems. Jour. Math. Anal. 54, 309–315 (1976)

[R5.44]   Weiss, M.D.: Fixed Points, Separation and Induced Topologies for Fuzzy Sets. Jour. Math. Anal. and Appln. 50, 142–150 (1975)

[R5.45]   Wilkinson, J.: Archetypes, Language, Dynamic Programming and Fuzzy Sets. In: Wilkinson, J., et al. (eds.) The Dynamic Programming of Human Systems, pp. 44–53. Information Corp., MSS New York (1973)

[R5.46]   Zadeh, L.A.: The Role of Fuzzy Logic in the Management of Ucertainty in expert Systems. Fuzzy Sets and Systems 11, 199–227 (1983)

[R5.47]   Zimmerman, H.-J.: Description and Optimization of Fuzzy Systems. Intern. Jour. Gen. Syst. 2(4), 209–215 (1975)

[R5.48]   Zimmerman, H.J.: Applications of Fuzzy Set Theory to Mathematical Programming. Information Science 36(1), 29–58 (1985)

## R6.  Fuzzy Probability, Fuzzy Random Variable and Random Fuzzy Variable

[R6.1]   Bandemer, H.: From Fuzzy Data to Functional Relations. Mathematical Modelling 6, 419–426 (1987)

[R6.2]   Bandemer, H., et al.: Fuzzy Data Analysis. Kluwer, Boston (1992)

[R6.3]   Kruse, R., et al.: Statistics with Vague Data. D. Reidel Pub. Co., Dordrecht (1987)

[R6.4]   Chang, R.L.P., et al.: Applications of Fuzzy Sets in Curve Fitting. Fuzzy Sets and Systems 2(1), 67–74

[R6.5]   Chen, S.Q.: Analysis for Multiple Fuzzy Regression. Fuzzy Sets and Systems 25(1), 56–65

[R6.6]      Celmins, A.: Multidimensional Least-Squares Fitting of Fuzzy Model. Mathematical Modelling 9(9), 669–690

[R6.7]      Dumitrescu, D.: Entropy of a Fuzzy Process. Fuzzy Sets and Systems 55(2), 169–177 (1993)

[R6.8]      Delgado, M., et al.: On the Concept of Possibility-Probability Consistency. Fuzzy Sets and Systems 21(3), 311–318 (1987)

[R6.9]      Devi, B.B., et al.: Estimation of Fuzzy Memberships from Histograms. Information Sciences 35(1), 43–59 (1985)

[R6.10]     Diamond, P.: Fuzzy Least Squares. Information Sciences 46(3), 141–157 (1988)

[R6.11]     Dubois, D., et al.: Fuzzy Sets, Probability and Measurement. European Jour. of Operational Research 40(2), 135–154 (1989)

[R6.12]     El Rayes, A.B., et al.: Generalized Possibility Measures. Information Sciences 79, 201–222 (1994)

[R6.13]     Fruhwirth-Schnatter, S.: On Statistical Inference for Fuzzy Data with Applications to Descriptive Statistics. Fuzzy Sets and Systems 50(2), 143–165 (1992)

[R6.14]     Fruhwirth-Schnatter, S.: On Fuzzy Bayesian Inference. Fuzzy Sets and Systems 60(1), 41–58 (1993)

[R6.15]     Gaines, B.R.: Fuzzy and Probability Uncertainty logics. Information and Control 38(2), 154–169 (1978)

[R6.16]     Geer, J.F., et al.: Discord in Possibility Theory. International Jour. of General Systems 19, 119–132 (1991)

[R6.17]     Geer, J.F., et al.: A Mathematical Analysis of Information-Processing Transformation Between Probabilistic and Possibilistic Formulation of Uncertainty. International Jour. of General Systems 20(2), 14–176 (1992)

[R6.18]     Goodman, I.R., et al.: Uncertainty Models for Knowledge Based Systems. North-Holland, New York (1985)

[R6.19]     Grabish, M., et al.: Fundamentals of Uncertainty Calculi with Application to Fuzzy Systems. Kluwer, Boston (1994)

[R6.20]     Guan, J.W., et al.: Evidence Theory and Its Applications, vol. 1. North-Holland, New York (1991)

[R6.21]     Guan, J.W., et al.: Evidence Theory and Its Applications, vol. 2. North- Holland, New York (1992)

[R6.22]     Hisdal, E.: Are Grades of Membership Probabilities? Fuzzy Sets and Systems 25(3), 349–356 (1988)

[R6.23]     Ulrich, H.: A Mathematical Theory of Uncertainty. In: Yager, R.R. (ed.) Fuzzy Set and Possibility Theory: Recent Developments, pp. 344–355. Pergamon, New York (1982)

[R6.24]     Kacprzyk, J., Fedrizzi, M. (eds.): Combining Fuzzy Imprecision with Probabilistic Uncertainty in Decision Making. Plenum Press, New York (1992)

[R6.25]     Kacprzyk, J., et al.: Combining Fuzzy Imprecision with Probabilistic Uncertainty in Decision Making. Springer, New York (1988)

[R6.26]     Klir, G.J.: Where Do we Stand on Measures of Uncertainty, Ambignity, Fuzziness and the like? Fuzzy Sets and Systems 24(2), 141–160 (1987)

[R6.27]     Klir, G.J., et al.: Fuzzy Sets, Uncertainty and Information. Prentice Hll, Englewood Cliff (1988)

[R6.28]     Klir, G.J., et al.: Probability-Possibility Transformations: A Comparison. Intern. Jour. of General Systems 21(3), 291–310 (1992)

[R6.29] Kosko, B.: Fuzziness vs Probability. Intern. Jour. of General Systems 17(1-3), 211–240 (1990)

[R6.30] Manton, K.G., et al.: Statistical Applications Using Fuzzy Sets. John Wiley, New York (1994)

[R6.31] Meier, W., et al.: Fuzzy Data Analysis: Methods and Indistrial Applications. Fuzzy Sets and Systems 61(1), 19–28 (1994)

[R6.32] Nakamura, A., et al.: A logic for Fuzzy Data Analysis. Fuzzy Sets and Systems 39(2), 127–132 (1991)

[R6.33] Negoita, C.V., et al.: Simulation, Knowledge-Based Compting and Fuzzy Statistics. Van Nostrand Reinhold, New York (1987)

[R6.34] Nguyen, H.T.: Random Sets and Belief Functions. Jour. of Math. Analysis and Applications 65(3), 531–542 (1978)

[R6.35] Prade, H., et al.: Representation and Combination of Uncertainty with belief Functions and Possibility Measures. Comput. Intell. 4, 244–264 (1988)

[R6.36] Puri, M.L., et al.: Fuzzy Random Variables. Jour. of Mathematical Analysis and Applications 114(2), 409–422 (1986)

[R6.37] Rao, N.B., Rashed, A.: Some Comments on Fuzzy Random Variables. Fuzzy Sets and Systems 6(3), 285–292 (1981)

[R6.38] Sakawa, M., et al.: Multiobjective Fuzzy linear Regression Analysis for Fuzzy Input-Output Data. Fuzzy Sets and Systems 47(2), 173–182 (1992)

[R6.39] Schneider, M., et al.: Properties of the Fuzzy Expected Values and the Fuzzy Expected Interval. Fuzzy Sets and Systems 26(3), 373–385 (1988)

[R6.40] Slowinski, R., Teghem, J. (eds.): Stochastic versus Fuzzy Approaches to Multiobjective Mathematical Programming Under Uncertainty. Kluwer, Dordrecht (1990)

[R6.41] Stein, N.E., Talaki, K.: Convex Fuzzy Random Variables. Fuzzy Sets and Systems 6(3), 271–284 (1981)

[R6.42] Sudkamp, T.: On Probability-Possibility Transformations. Fuzzy Sets and Systems 51(1), 73–82 (1992)

[R6.43] Tanaka, H., et al.: Possibilistic Linear Regression Analysis for Fuzzy Data. European Jour. of Operational Research 40(3), 389–396 (1989)

[R6.44] Walley, P.: Statistical Reasoning with Imprecise Probabilities. Chapman and Hall, London (1991)

[R6.45] Wang, G.Y., et al.: The Theory of Fuzzy Stochastic Processes. Fuzzy Sets and Systems 51(2), 161–178 (1992)

[R6.46] Wang, X., et al.: Fuzzy Linear Regression Analysis of Fuzzy Valued Variable. Fuzzy Sets and Systems 36(1)

[R6.47] Zadeh, L.A.: Probability Measure of Fuzzy Event. Jour. of Math Analysis and Applications 23, 421–427 (1968)

## R7. Ideology and the Knowledge Construction Process

[R7.1] Abercrombie, N., et al.: The Dominant Ideology Thesis. Allen and Unwin, London (1980)

[R7.2] Abercrombie, N.: Class, Structure, and Knowledge: Problems in the Sociology of Knowledge. New York University Press, New York (1980)

[R7.3] Aron, R.: The Opium of the Intellectuals. University Press of America, Lanham (1985)

[R7.4]    Aronowitz, S.: Science as Power: Discourse and Ideology in Modern Society. University of Minnesota Press, Minneapolis (1988)

[R7.5]    Barinaga, M., Marshall, E.: Confusion on the Cutting Edge. Science 257, 616–625 (1992)

[R7.6]    Barnett, R.: Beyond All Reason: Living with Ideology in the University. Society for Research into Higher Education and Open University Press, Philadelphia (2003)

[R7.7]    Barth, H.: Truth and Ideology. University of California Press, Berkeley (1976)

[R7.8]    Basin, A., Verdie, T.: The Economics of Cultural Transmission and the Dynamics of Preferences. Journal of Economic Theory 97, 298–319 (2001)

[R7.9]    Beardsley, P.L.: Redefining Rigor: Ideology and Statistics in Political Inquiry. Sage Publications, Bevery Hills (1980)

[R7.10]   Bikhchandani, S., et al.: A Theory of Fads, Fashion, Custom, and Cultural Change. Journal of Political Economy 100, 992–1026 (1992)

[R7.11]   Robert, B., Richerson, P.J.: Culture and Evolutionary Process. University of Chicago Press, Chicago (1985)

[R7.12]   Buczkowski, P., Klawiter, A.: Theories of Ideology and Ideology of Theories. Rodopi, Amsterdam (1986)

[ R7.13]  Chomsky, N.: Manufacturing Consent. Pantheo Press, New York (1988)

[R7.14]   Chomsky, N.: Problem of Knowledge and Freedom. Collins, Glasgow (1972)

[R7.15]   Cole, J.R.: Patterns of Intellectual influence in Scientific Research. Sociology of Education 43, 377–403 (1968)

[R7.16]   Cole Jonathan, R., Cole, S.: Social Stratification in Science. University of Chicago Press, Chicago (1973)

[R7.17]   Debackere, K., Rappa, M.A.: Institutioal Varations in Problem Choice and Persistence among Scientists in an Emerging Fields. Research Policy 23, 425–441 (1994)

[R7.18]   Fraser, C., Gaskell, G.: The Social Psychological Study of Widespread Beliefs. Clarendon Press, Oxford (1990)

[R7.19]   Gieryn, T.F.: Problem Retention and Problem Change in Science. Sociological Inquiry 48, 96–115 (1978)

[R7.20]   Harrington Jr., J.E.: The Rigidity of social Systems. Journal of Political Economy 107, 40–64

[R7.21]   Hinich, M., Munger, M.: Ideology and the Theory of Political Choice. University of Michigan Press, Ann Arbor (1994)

[R7.22]   Hull, D.L.: Science as a Process: An Evolutionary Account of the Social and Conceptual Development of Science. University of Chicago Press, Chicago (1988)

[R7.23]   Mészáros, I.: Philosophy, Ideology and Social Science: Essay in Negation and Affirmation, Brighton Wheatsheaf, Sussex (1986)

[R7.24]   Mészáros, I.: The Power of Ideology. New York University Press, New York (1989)

[R7.25]   Newcomb, T.M., et al.: Persistence and Change. John Wiley, New York (1967)

[R7.26]   Pickering, A.: Science as Practice and Culture. University of Chicago Press, Chicago (1992)

[R7.27]   Therborn, G.: The Ideology of Power and the Power of Ideology. NLB Publications, London (1980)

[R7.28]   Thompson, K.: Beliefs and Ideology. Tavistock Publication, New York (1986)

[R7.29]   Ziman, J.: The Problem of 'Problem Choice'. Minerva 25, 92–105 (1987)

[R7.30] Ziman, J.: Public Knowledge: An Essay Concerning the Social Dimension of Science. Cambridge University Press, Cambridge (1968)

[R7.31] Zuckerman, H.: Theory Choice and Problem Choice in Science. Sociological Inquiry 48, 65–95 (1978)

# R 8. Information, Thought and Knowledge

[R8.1] Aczel, J., Daroczy, Z.: On Measures of Information and their Characterizations. Academic Press, New York (1975)

[R8.2] Angelov, S., Georgiev, D.: The Problem of Human Being in Contemporary Scientific Knowledge. Soviet Studies in Philosophy, 49–66 (Summer 1974)

[R8.3] Anderson, J.R.: The Architecture of Cognition. Harvard University Press, Cambridge (1983)

[R8.4] Angelov, S., Georgiev, D.: The Problem of Human Being in Contemporary Scientic Knowledge. Soviet Studies in Philosophy, 49–66 (Summer 1974)

[R8.5] Ash, R.: Information Theory. John Wiley and Sons, New York (1965)

[R8.6] Barlas, Y., Carpenter, S.: Philosophical Roots of Model Validation: Two Paradigms. System Dynamic Review 6, 148–166 (1990)

[R8.7] Barrett, T.W.: Quantum Statistical Foundations for Structural Information Theory and Communication Theory. In: Lakshmikantham, V. (ed.) Nonlinear Systems and Applications, pp. 389–407. Academic Press, New York (1977)

[R8.8] Bergin, J.: Common Knowledge with Monotone Statistics. Econometrica 69, 1315–1332 (2001)

[R8.9] Bestougeff, H., Ligozat, G.: Logical Tools for Temporal Knowledge Representation. Ellis Horwood, New York (1992)

[R8.10] Brillouin, L.: Science and information Theory. Academic Press, New York (1962)

[R8.11] Bruner, J.S., et al.: A Study of Thinking. Wiley, New York (1956)

[R8.12] Brunner, K., Meltzer, A.H. (eds.): Three Aspects of Policy and Policy Making: Knowledge, Data and Institutions. Carnegie-Rochester Conference Series, vol. 10. North-Holland, Amsterdam (1979)

[R8.13] Burks, A.W.: Chance, Cause, Reason: An Inquiry into the Nature of Scientific Evidence. University of Chicago Press, Chicago (1977)

[R8.14] Calvert, R.: Models of Imperfect Information in Politics. Hardwood Academic Publishers, New York (1986)

[R8.15] Cornforth, M.: The Theory of Knowledge. International Pub., New York (1972)

[R8.16] Coombs, C.H.: A Theory of Data. Wiley, New York (1964)

[R8.17] Crawshay-Willims: Methods and Criteria of Reasoning. Routledge and Kegan Paul, London (1957)

[R8.18] Dretske, F.I.: Knowledge and the Flow of Information. MIT Press, Cambridge (1981)

[R8.19] Dreyfus, H.L.: A Framework for Misrepresenting Knowledge. In: Ringle, M. (ed.) Philosophical Perspectives in Artificial Intelligence. Humanities Press, Atlantic Highlands (1979)

[R8.20] Fagin, R., Halpenn, J.Y.: Reasoning About Knowledge and Probability. In: Vardi, M. (ed.) Proceedings of the Second Conference of Theoretical Aspects of Reasoning About Knowledge, pp. 277–293. Morgan Kaufmann, Asiloman (1988)

[R8.21a]  Fagin, R., et al.: Reasoning About Knowledge. MIT Press, Cambridge (1995)

[R8.21b]  Fedoseyev, P.N.: Scientific Cognition Today, Its Specific Features and Problems. In: Philosophy in the USSR: Problems of Dialectical Materialism (1977) (Translated by Robert Daglish, Moscow, Progress Publishers)

[R8.22]  Geanakoplos, J.: Common Knowledge. In: Moses, Y. (ed.) Proceedings of the Fourth Conference of Theoretical Aspects of Reasoning About Knowledge. Morgan Kaufmann, San Mateo (1992)

[R8.23]  Geanakoplos, J.: Common Knowledge. Journal of Economic Perspectives 6, 53–82 (1992)

[R8.24]  George, F.H.: Models of Thinking. Allen and Unwin, London (1970)

[R8.25]  George, F.H.: Epistemology and the problem of perception. Mind 66, 491–506 (1957)

[R8.26]  Harwood, E.C.: Reconstruction of Economics. American Institute for Economic Research, Great Barrington (1955)

[R8.27]  Hintikka, J.: Knowledge and Belief. Cornell University Press, Ithaca (1962)

[R8.28]  Hirshleifer, J.: The Private and Social Value of Information and Reward to inventive activity. American Economic Review 61, 561–574 (1971)

[R8.29]  Hirshleifer, J., Riley, J.: The Analytics of Uncertainty and Information: An expository Survey. Journal of Economic Literature 17, 1375–1421 (1979)

[R8.30]  Hirshleifer, J., Riley, J.: The Economics of Uncertainty and Information. Cambridge University Press, Cambridge (1992)

[R8.31]  Kapitsa, P.L.: The Influence of Scientific Ideas on Society, pp. 52–71. Soviet Studies in Philosophy (Fall 1979)

[R8.32]  Kedrov, B.M.: The Road to Truth. Soviet Studies in Philosophy 4, 3–53 (1965)

[R8.33]  Klatzky, R.L.: Human Memory: Structure and Processes. W. H. Freeman Pub., San Francisco (1975)

[R8.34]  Koopmans, T.C.: Three Essays on the State of Economic Science. McGraw-Hill, New York (1957)

[R8.35]  Kreps, D., Wilson, R.: Reputation and Imperfect Information. Journal of Economic Theory 27, 253–279 (1982)

[R8.36]  Kubat, L., Zeman, J.: Entropy and Information. Elsevier, Amsterdam (1975)

[R8.37]  Kurcz, G., Shugar, W., et al. (eds.): Knowledge and Language. North-Holland, Amsterdam (1986)

[R8.38]  Lakemeyer, G., Nobel, B.: Foundations of Knowledge Representation and Reasoning. Springer, Berlin (1994)

[R8.39]  Lektorskii, V.A.: Principles involved in the Reproduction of Objective in Knowledge. Soviet Studies in Philosophy 4(4), 11–21 (1967)

[R8.40]  Levi, I.: The Enterprise of Knowledge. MIT Press, Cambridge (1980)

[R8.41]  Levi, I.: Ignorance, Probability and Rational Choice. Synthese 53, 387–417 (1982)

[R8.42]  Levi, I.: Four Types of Ignorance. Social Science 44, 745–756

[R8.43]  Levine, D., Aparicio IV, M.: Neural Networks for Knowledge Representation and Inference. Lawrence Erlbaum Associates Publishers, Hillsdale (1994)

[R8.44]  Marschak, J.: Economic Information, Decision and Prediction: Selected Essays, Part II, vol. II. Dordrecnt-Holland, Boston (1974)

[R8.45]  McDermott, J.: Representing Knowledge in Intelligent Systems. In: Ringle, M. (ed.) Philosophical Perspectives in Artificial Intelligence, pp. 110–123. Humanities press, Atlantic Highlands (1979)

[R8.46]   Menges, G. (ed.): Information, Inference and Decision. D. Reidel Pub., Dordrecht (1974)

[R8.47]   Masuch, M., Pólos, L. (eds.): Knowledge Representation and Reasoning Under Uncertainty. Springer, New York (1994)

[R8.48]   Moses, Y. (ed.): Proceedings of the Fourth Conference of Theoretical Aspects of Reasoning about Knowledge. Morgan Kaufmann, San Mateo (1992)

[R8.49]   Nielsen, L.T., et al.: Common Knowledge of Aggregation Expectations. Econometrica 58, 1235–1239 (1990)

[R8.50]   Newell, A.: Unified Theories of Cognition. Harvard University Press, Cambridge (1990)

[R8.51]   Newell, A.: Human Problem Solving. Prentice-Hall, Englewood Cliff (1972)

[R8.52]   Ogden, G.K., Richards, I.A.: The Meaning of Meaning. Harcourt-Brace Jovanovich, New York (1923)

[R8.53]   Planck, M.: Scientific Autobiography and Other Papers. Westport, Conn., Greenwood (1968)

[R8.54]   Pollock, J.: Knowledge and Justification. Princeton University Press, Princeton (1974)

[R8.55]   Polanyi, M.: Personal Knowledge. Routledge and Kegan Paul, London (1958)

[R8.56]   Popper, K.R.: Objective Knowledge. Macmillan, London (1949)

[R8.57]   Price, H.H.: Thinking and Experience. Hutchinson, London (1953)

[R8.58]   Putman, H.: Reason, Truth and History. Cambridge University Press, Cambridge (1981)

[R8.59]   Putman, H.: Realism and Reason. Cambridge University Press, Cambridge (1983)

[R8.60]   Putman, H.: The Many Faces of Realism. Open Court Publishing Co., La Salle (1987)

[R8.61]   Rothschild, K.W.: Models of Market Organization with Imperfect Information: A Survey. Journal of Political Economy 81, 1283–1308 (1973)

[R8.62]   Russell, B.: Human Knowledge, its Scope and Limits. Allen and Unwin, London (1948)

[R8.63]   Russell, B.: Our Knowledge of the External World. Norton, New York (1929)

[R8.64]   Samet, D.: Ignoring Ignorance and Agreeing to Disagree. Journal of Economic Theory 52, 190–207 (1990)

[R8.65]   Schroder, H.M., Suedfeld, P. (eds.): Personality Theory and Information Processing. Ronald Pub., New York (1971)

[R8.66]   Searle, J.: Minds, Brains and Science. Harvard University Press, Cambridge (1985)

[R8.67]   Sen, A.K.: On Weights and Measures: Information Constraints in Social Welfare Analysis. Econometrica 45, 1539–1572 (1977)

[R8.68]   Shin, H.: Logical Structure of Common Knowledge. Journal of Economic Theory 60, 1–13 (1993)

[R8.69]   Simon, H.A.: Models of Thought. Yale University Press, New Haven (1979)

[R8.70]   Smithson, M.: Ignorance and Uncertainty, Emerging Paradigms. Springer, New York (1989)

[R8.71]   Sowa, J.F.: Knowledge Representation: Logical, Philosophical and Computational Foundations. Brooks Pub., Pacific Grove (2000)

[R8.72]   Stigler, G.J.: The Economics of Information. Journal of Political Economy 69, 213–225 (1961)

[R8.73]   Tiukhtin, V.S.: How Reality Can be Reflected in Cognition: Reflection as a
          Property of All Matter. Soviet Studies in Philosophy 3(1), 3–12 (1964)
[R8.74]   Tsypkin, Y.Z.: Foundations of the Theory of Learning Systems. Academic
          Press, New York (1973)
[R8.75]   Ursul, A.D.: The Problem of the Objectivity of Information. In: Kubát, L.,
          Zeman, J. (eds.) Entropy and Information, pp. 187–230. Elsevier, Amsterdam
          (1975)
[R8.76]   Vardi, M. (ed.): Proceedings of Second Conference on Theoretical Aspects of
          Reasoning about Knowledge. Morgan Kaufman, Los Altos (1988)
[R8.77]   Vazquez, M., et al.: Knowledge and Reality: Some Conceptual Issues in System
          Dynamics Modeling. Systems Dynamics Review 12, 21–37 (1996)
[R8.78]   Zadeh, L.A.: A Theory of Commonsense Knowledge. In: Skala, H.J., et al. (eds.)
          Aspects of Vagueness, pp. 257–295. D. Reidel Co., Dordrecht (1984)
[R8.79]   Zadeh, L.A.: The Concept of Linguistic Variable and its Application to
          Approximate reasoning. Information Science 8, 40–80 (1975) (Also in Vol. 9,
          pp. 40 – 80)

# R9.  Language and the Knowledge-Production Process

[R9.1]    Aho, A.V.: Indexed Grammar - An Extension of Context-Free Grammars.
          Journal of the Association for Computing Machinery 15, 647–671 (1968)
[R9.2]    Black, M. (ed.): The Importance of Language. Prentice- Hall, Englewood Cliffs
          (1962)
[R9.3]    Buchler, J.: Metaphysics of Natural Complexes. Columbia University Press,
          New York (1966)
[R9.4]    Carnap, R.: Meaning and Necessity: A Study in Semantics and Modal Logic.
          University of Chicago Press, Chicago (1956)
[R9.5]    Chomsky, N.: Linguistics and Philosophy. In: Hook, S. (ed.) Language and
          Philosophy, pp. 51–94. New York University Press, New York (1968)
[R9.6]    Chomsky, N.: Language and Mind. Harcourt Brace Jovanovich, New York
          (1972)
[R9.7]    Cooper, W.S.: Foundations of Logico-Linguistics: A Unified Theory of
          Information, Language and Logic. D. Reidel, Dordrecht (1978)
[R9.8]    Cresswell, M.J.: Logics and Languages. Methuen Pub., London (1973)
[R9.9]    Dilman, I.: Studies in Language and Reason. Barnes and Nobles, Books, Totowa
          (1981)
[R9.10]   Fodor, J.A.: The Language and Thought. Thomas Y. Crowell Co., New York
          (1975)
[R9.11]   Ginsbury, S.: Algebraic and Automata – Theoretical properties of Formal
          Languages. North-Holland, New York (1975)
[R9.12]   Givon, T.: On Understanding Grammar. Academic Press, New York (1979)
[R9.13]   Gorsky, D.R.: Definition. Progress Publishers, Moscow (1974)
[R9.14]   Greibach, S.A.: An Infinite Hierarchy of Contex-Free Languages. Journal of
          Associazion for Computing Machinery 16, 91–106 (1969)
[R9.15]   Hintikka, J.: The Game of Language. D. Reidel Pub., Dordrecht (1983)
[R9.16]   Johnson-Lair, P.N.: Mental Models: Toward Cognitive Science of Language,
          Inference and Consciousness. Harvard University Press, Cambridge (1983)

[R9.17]   Kandel, A.: Codes Over Languages. IEEE Transactions on Systems Man and Cybernetics 4, 135–138 (1975)

[R9.18]   Keenan, E.L., Faltz, L.M.: Boolean Semantics for Natural Languages. D. Reidel Pub., Dordrecht

[R9.19]   Lakoff, G.: Linguistics and Natural Logic. Synthese 22, 151–271 (1970)

[R9.20]   Lee, E.T., et al.: Notes On Fuzzy Languages. Information Science 1, 421–434 (1969)

[R9.21]   Mackey, A., Merrill, D. (eds.): Issues in the Philosophy of Language. CT. Yale University Press, New Haven (1976)

[R9.22]   Nagel, T.: Linguistics and Epistemology. In: Hook, S. (ed.) Language and Philosophy, pp. 180–184. New York University Press, New York (1969)

[R9.23]   Pike, K.: Language in Relation to a Unified Theory of Structure of Human Behavior. Mouton Pub., The Hague (1969)

[R9.24]   Quine, W.V.O.: Word and object. MIT Press, Cambridge (1960)

[R9.25]   Russell, B.: An Inquiry into Meaning and Truth. Penguin Books (1970)

[R9.26]   Salomaa, A.: Formal Languages. Academic Press, New York (1978)

[R9.27]   Tamura, S., et al.: Learning of Formal Language. IEEE Trans. Syst. Man. Cybernetics, SMC 3, 98–102 (1973)

[R9.28]   Tarski, A.: Logic, Semantics and Matamathematics. Clarendon Press, Oxford (1956)

[R9.29]   Whorf, B.L. (ed.): Language, Thought and Reality. Humanities Press, New York (1956)

[R9.30]   Winogrand, T.: Understanding Natural Language. Cognitive Psychology 3, 1–191 (1972)

# R10. Probabilistic Concepts and Reasoning

[R10.1]   Anscombe, F., Aumann, R.J.: A Definition of Subjective Probability. Annals of Mathematical Statistics 34, 199–205 (1963)

[R10.2]   Billingsley, P.: Probability and Measure. John Wiley and Sons, New York (1979)

[R10.3]   Boolos, G.S., Jeffrey, R.C.: Computability and Logic. Cambridge University Press, New York (1989)

[R10.4]   Carnap, R.: Logical Foundation of Probability. Routledge and Kegan Paul Ltd., London (1950)

[R10.5]   Cohen, L.J.: The Probable and Provable. Clarendon Press, Oxford (1977)

[R10.6]   de Finetti, B.: Probabilities of Probabilities a Real Problem or Misunderstanding? In: Aykac, A., et al. (eds.) New Developments in the Applications of Bayesian Methods, Amsterdam, pp. 1–10 (1977)

[R10.7]   Dempster, A.P.: Upper and Lower Probabilities Induced by Multivalued Mapping. Annals of Math Statistics 38, 325–339 (1967)

[R10.8]   Domotor, Z.: Higher Order Probabilities. Philosophical Studies 40, 31–46 (1981)

[R10.9]   Doob, J.L.: Stochastic Processes. John Wiley and Sons, Chichester (1990)

[R10.10]  Fellner, W.: Distortion of Subjective Probabilities as a Reaction to Uncertainty. Quarterly Journal of Economics 75, 670–689 (1961)

[R10.11]  Fishburn, P.C.: The Axioms of Subjective Probability. Statistical Sciences 1(3), 335–358 (1986)

[R10.12]  Fishburn, P.C.: Decision and Value. John Wiley and Sons, New York (1964)

[R10.13]  Gaifman, C.: A Theory of Higher Order Probabilities. In: Halpern, J.Y. (ed.) Theoretical Aspects of Reasoning about Knowledge, pp. 275–292. Morgan Kaufman, Los Alamitos (1986)

[R10.14]  George, F.H.: Logical Networks and Probability. Bulletin of Mathematical Biophysics 19, 187–199 (1957)

[R10.15]  Good, I.J.: Probability and the Weighing of Evidence. Charles Griffin and Co. Ltd., London (1950)

[R10.16]  Good, I.J.: Subjective Probability as the Measure of Non-measurable Set. In: Nagel, E., et al. (eds.) Logic, Methodology, and the Philosophy of Science, pp. 319–329. Stanford University Press, Stanford (1962)

[R10.17]  Good, I.J.: Good Thinking: The Foundations of Probability and Applications. University of Minnesota Press, Minneapolis (1983)

[R10.18]  Goutsias, J., et al. (eds.): Random Sets: Theory and Applications. Springer, New York (1997)

[R10.19]  Hacking, I.: The emergence of Probability. Cambridge University Press, London (1975)

[R10.20]  Harsanyi, J.C.: Acceptance of Empirical Statements: A Bayesian Theory without Cognitive Utilities. Theory and Decision 18, 1–30 (1985)

[R10.21]  Ulrich, H., Klement, E.P.: Plausibility Measures: A General Framework for Possibility and Fuzzy Probability Measures. In: Skala, H.J., et al. (eds.) Aspects of Vagueness, pp. 31–50. D. Reidel Pub. Co., Dordrecht (1984)

[R10.22]  Holmos, P.R.: Measure Theory. Van Nostrand, New York (1950)

[R10.23]  Hoover, D.N.: Probabilistic Logic. Annals of Mathematical Logic 14, 287–313 (1978)

[R10.24]  Jeffery, R.: The Present Position in Probability Theory. British Journal for the Philosophy of Science 5, 275–280 (1955)

[R10.25]  Kahneman, D., Tversky, A.: Sujective Probability: A Judgment of representativeness. Cognitive Psychology 3, 430–454 (1972)

[R10.26]  Keynes, J.M.: A treatise on Probability. MacMillan and Co., London (1921)

[R10.27]  Kolmogrov, A.N.: Foundation of the Theory of Probability. Chelsea Pub. Co., New York (1956)

[R10.28]  Koopman, B.O.: The Axioms and Algebra of Intuitive Probability. Annals of Mathematics 41, 269–292 (1940)

[R10.29]  Kraft, C., et al.: Intuitive Probability on Finite Sets. Annals of Mathematical Statistics 30, 408–419 (1959)

[R10.30]  Kullback, S., Leibler, R.A.: Information and Sufficiency. Annals of Math. Statistics 22, 79–86 (1951)

[R10.31]  Kyburg, H.E.: Probability and the Logic of Rational Belief. Wesleyan University Press, Middleton (1961)

[R10.32]  Kyburg, H.E., Smokler, H.E.: Studies in Subjective Probability. Wiley, New York (1964)

[R10.33]  Laha, R., Rohatgi, V.K.: Probability Theory. John Wiley and Sons, New York (1979)

[R10.34]  Laplace, P.S.: A Philosophical Essay on Probabilities. Constable and Co., London (1951)

[R10.35]  Matheron, G.: Random Sets and Integral Geometry. John Wiley and Sons, New York (1975)

[R10.36] Marschak, J.: Personal Probabilities of Probabilities. Theory and Decision 6, 121–153 (1975)

[R10.37] Nagel, E.: Principles of the Theory of Probability. In: Neurath, O., et al. (eds.) International Encyclopedia of Unified Science, vol. 1-10, pp. 343–422. University of Chicago Press, Chicago (1955)

[R10.38] Nilsson, N.J.: Probabilistic Logic. Probabilistic Logic 28, 71–87 (1986)

[R10.39] Parrat, L.G.: Probability and Experimental Errors in Science. John Wiley and Sons, New York (1961)

[R10.40] Patty Wayne, C.: Foundations of Topology. PWS Pub. Co., Boston (1993)

[R10.41] Parthasarath, K.R.: Probability Measure on Metric Spaces. Academic Press, New York (1967)

[R10.42] Ruspini, E.: Epistemic Logics, Probability and Calculus of Evidence. In: Proceedings of 10th International Joint Conference on AI (IJCAI 1987), Milan, pp. 924–931 (1987)

[R10.43] Savage, L.J.: The Foundations of Statistics. John Wiley and Sons, New York (1954)

[R10.44] Schneeweiss, H.: Probability and Utility – Dual Concepts in Decision Theory I. In: Menges (ed.) Information, Inference and Decision. Reidel, Dordrecht (1974)

[R10.45] Shafer, G.: A Mathematical Theory of Evidence. Princeton University Press, Princeton (1976)

[R10.46] Shafer, G.: Constructive Probability. Synthese 48, 1–60 (1981)

[R10.47] Shannon, C.E., Weaver, W.: The Mathematical Theory of Communication. University of Illinois Press, Urbana (1949)

[R10.48] Tiller, P., Green, E.D. (eds.): Probability and Inference in the Law of Evidence: The Uses and Limits of Bayesianism. Kluwer Academic Publishers, Dordrecht (1988)

[R10.49] von Mises, R.: Probability, Statistics and Truth. Dover Pub., New York (1981)

[R10.50] Wagon, S.: The Banach-Tarski Paradox. Cambridge University Press, Cambridge (1985)

# R11. Optimality, Classical Exactness and Equilibrium in Knowledge Systems

[R11.1] Agassi, J., Jarvie, I.C. (eds.): Rationality: The Critical View. M. Nijhoff Pub., Boston (1987)

[R11.2] Amsterdamski, S.: Between History and Method: Disputes about the Rationality of Science. Kluwer Academic Pub., Dordrecht (1992)

[R11.3] Anderson, G.: Rationality in Science and Politics. D. Reid Pub. Co., Dordrecht (1984)

[R11.4] Arrow, K.J.: Rationality of self and Others in an Economic System. Journal of Business 59, 385–399 (1986)

[R11.5] Baumol, W., Quant, R.: Rules of Thumb and Optimally Imperfect Decisions. American Economic Review 54, 23–46 (1964)

[R11.6] Benn, S.I., Mortimore, G.W. (eds.): Rationality and the Social Sciences: Contributions to the Philosophy and Methodology of the Social Sciences. Routledge and Kegan Paul, London (1976)

[R11.7]   Bicchieri, C.: Rationality and Coordination. Cambridge University Press, New York (1993)

[R11.8]   Biderman, S., Scharfstein, B.-A.: Rationality in a question: East and Western View of Rationality. E.J. Brill Pub., New York (1989)

[R11.9]   Black, F.: Exploring General Equilibrium. MIT Press, Cambridge (1995)

[R11.10]  Boland, L.A.: On the Futility of Criticizing the Neoclassical Maximization Hypothesis. The American Economic Review 71, 1031–1036 (1981)

[R11.11]  Border, K.C.: Fixed Point Theorems with Applications to Economics and Game Theory. Cambridge University Press, Cambridge (1985)

[R11.12]  Bowman, E.H.: Consistency and Optimality in Managerial Decision Making. Management Science 9, 310–321 (1963)

[R11.13]  Brubaker, R.: The Limits of Rationality: An Essay on the Social and Moral Thought of Max Weber. Allen and Unwin, London (1984)

[R11.14]  Churchman, C.W.: Prediction and Optimal Decision. Prentice-Hall, Englewood Cliffs (1961)

[R11.15]  Cohen, L.J.: Can Human Irrationality be Experimentally Demonstrated? Behavioral and Brain Science 4, 317–370 (1981)

[R11.16]  Cohen, M., Jaffray, J.-Y.: Is Savage's Independence Axiom a Universal Rationality Principle? Behavioral Science 33, 38–47

[R11.17]  Cornwall, R.R.: Introduction to the use of General Equilibrium Analysis. North-Holland, New York (1984)

[R11.18]  De Sousa, R.: The Rationality of Motion. MIT Press, Cambridge

[R11.19]  Dompere, K.K.: Fuzziness, Rationality, Optimality and Equilibrium in Decision and Economic Theories. In: Lodwick, W.A., Kacprzyk, J. (eds.) Fuzzy Optimization. STUDFUZZ, vol. 254, pp. 3–32. Springer, Heidelberg (2010)

[R11.20]  Dompere, K.K.: On Epistemology and Decision-Choice Rationality. In: Trappl, R. (ed.) Cybernetics and System Research, pp. 219–228. North-Holland, New York (1982)

[R11.21]  Dompere, K.K.: Epistemic Foundations of Fuzziness: Unified Theories of Decision-Choice Processes. STUDFUZZ, vol. 236. Springer, New York (2009)

[R11.22]  Dompere, K.K.: Fuzziness and Approximate Reasoning: Epistemics on Uncertainty, Expectations and Risk in Rational Behavior. STUDFUZZ, vol. 237. Springer, New York (2009)

[R11.23]  Elster, J.: Ulysses and the Sirens: Studies in Rationality and Irrationality. Cambridge University Press, New York (1979)

[R11.24]  Elster, J.: Studies in the Subversion of Rationality. Cambridge University Press, New York (1983)

[R11.25]  Ernst, G.C., et al.: Principles of Structural Equilibrium, a Study of Equilibrium Conditions by Graphic, Force Moment and Virtual Displacement (virtual work). University of Nebraska Press, Lincoln (1962)

[R11.26]  Fischer, R.B., Peters, G.: Chemical Equilibrium. Saunders Pub., Philadelphia (1970)

[R11.27]  Fisher, F.M.: Disequilibrium Foundations of Equilibrium Economics. Cambridge University Press, New York (1983)

[R11.28]  Fourgeaud, C., Gouriéroux, C.: Learning Procedures and Convergence to Rationality. Econometrica 54, 845–868 (1986)

[R11.29]  Freeman, A., Carchedi, G. (eds.): Marx and Non-Equilibrium Economics. Edward Elgar, Cheltenham (1996)

[R11.30]  Newton, G., Hare, P.H. (eds.): Naturalism and Rationality. Prometheus Books, Buffalo (1986)

[R11.31]  Ginsburgh, V.: Activity Analysis and General Equilibrium Modelling. North-Holland, New York (1981)

[R11.32]  Hahn, F.: Equilibrium and Macroeconomics. MIT Press, Cambridge (1984)

[R11.33]  Hansen, B.: A Survey of General Equilibrium Systems. McGraw-Hill, New York (1970)

[R11.34]  Hilpinen, R. (ed.): Rationality in Science: Studies in the Foundations of Science and Ethics. D. Reidel, Dordrecht (1980)

[R11.35]  Hogarth, R.M., Reder, M.W. (eds.): Rational Choice: The Contrast between Economics and Psychology. University of Chicago Press, Chicago (1986)

[R11.36]  Hollis, M., Lukes, S.: Rationality and Relativism. MIT Press, Cambridge (1982)

[R11.37]  Martin, H., Nell, E.J.: Rational Economic Man: A Philosophical Critique of Neo-classical Economics. Cambridge University Press, New York (1975)

[R11.38]  Howard, N.: Paradoxes of Rationality. MIT Press, Cambridge (1973)

[R11.39]  Hurwicz, L.: Optimality Criteria for Decision Making Under Ignorance, Mimeographed, Cowles Commission Discussion Paper. Statistics 370 (1951)

[R11.40]  Istratescu, V.I.: Fixed Point Theory: Introduction. D. Reidel Pub. Co, Dordrecht (1981)

[R11.41]  Keita, L.D.: Science, Rationality and Neoclassical Economics. University of Delaware Press, Newark (1992)

[R11.42]  Kirman, A. (ed.): Elements of General Equilibrium Analysis. Blackwell, Malden (1998)

[R11.43]  Kirman, A., Salmon, M. (eds.): Learning and Rationality in Economics. Basil Blackwell, Oxford (1993)

[R11.44]  Kornia, J.: Anti-Equilibrium. North-Holland, Amsterdam (1971)

[R11.45]  Kramer, G.H.: A Dynamic Model of Political Equilibrium. Journal of Economic Theory 16, 310–334 (1977)

[R11.46]  Kramer, G.: On a Class of Equilibrium Conditions for Majority Rule. Econometrica 41, 285–297 (1973)

[R11.47]  Kuenne, R.E.: The Theory of General Economic Equilibrium. Princeton University Press, Princeton (1967)

[R11.48]  Marschak, J.: Actual Versus Consistent Decision Behavior. Behavioral Science 9, 103–110 (1964)

[R11.49]  McKenzie, L.W.: Classical General Equilibrium Theory. MIT Press, Cambridge (2002)

[R11.50]  McMullin, E. (ed.): Construction and Constraint: The Shaping of Scientific Rationality. University of Notre Dame Press, Notre Dame Ind. (1988)

[R11.51]  Jozef, M.: The Problem of Rationality in Science and its Philosophy: Popper vs. Polanyi (The Polish Conferences). Kluwer Academic Publishers, Boston (1995)

[R11.52]  Mongin, P.: Does Optimization Imply Rationality. Synthese 124, 73–111 (2000)

[R11.53]  Negishi, T.: Microeconomic Foundations of Keynesian Macroeconomics. North-Holland, Amsterdam (1979)

[R11.54]  Newton-Smith, W.H.: The Rationality of Science. Routledge and Kegan Paul, Boston (1981)

[R11.55]  Page, S.E.: Two Measures of Difficulty. Economic Theory 8, 321–346 (1996)

[R11.56]  Parfit, D.: Personal Identity and Rationality. Synthese 53, 227–241 (1982)

[R11.57]  Pitt, J.C., Pera, M.: Rational Changes in Science: Essays on Scientific Reasoning. D. Reidel, Dordrecht (1987)

[R11.58]  Plott, C.R.: A Notion of Equilibrium and Its Possibility under Majority Rule. Amer. Econ. Rev. 57, 787–806 (1967)

[R11.59]  Preston, C.J.: Random Fields. Springer, Berlin (1976)

[R11.60]  Quirk, J., Saposnik, R.: Introduction to General Equilibrium Theory and Welfare Economics. McGraw-Hill, New York (1968)

[R11.61]  Radner, R.: Competitive Equilibrium Under Uncertainty. Econometrica 36, 31–58 (1968)

[R11.62]  Rapart, A.: Escape from Paradox. Scientific American, 50–56 (July 1967)

[R11.63]  Rescher, N.: Reason and Rationality in Natural Science: A Group of Essays. University Press of America, Lanham (1985)

[R11.64]  Sen, A.: Rational Behavior. In: Eatwell, J., et al. (eds.) Utility and Probability, pp. 198–216. Norton, New York (1990)

[R11.65]  Shackle, G.: Epistemics and Economics. Cambridge University Press, Cambridge (1973)

[R11.66]  Simon, H.A.: Rational Choice and the Structure of Environment. Psychological Review 63, 129–138 (1956)

[R11.67]  Simon, H.A.: Rationality as Process and as Product of Thought. American Economic Review 68, 1–16 (1978)

[R11.68]  Simon, H.A.: A Behavioral Model of Rational Choice. Quarterly Journal of Economics 69, 99–118 (1955)

[R11.69]  Simon, H.A.: Models of Man: Social and Rational. Wiley, New York (1957)

[R11.70]  Smart, D.R.: Fixed Point Theorems. Cambridge University Press, Cambridge (1980)

[R11.71]  Stambaugh, J.: The Real is not the Rational. State University of New York Press, Albany (1986)

[R11.72]  Tamny, M., Irani, K.D. (eds.): Rationality in Thought and Action. Greenwood Press, New York (1986)

[R11.73]  Turski, W.G.: Toward a Rationality of Emotions; Essay in the Philosophy of Mind. Ohio University Press, Athens (1994)

[R11.74]  Torr, C.: Equilibrium, Expectations, and Information: A Study of General Theory and Modern Classical Economics. Westview Press, Boulder Colo (1988)

[R11.75]  Valentinuzzi, M.: The Organs of Equilibrium and Orientation as a Control System. Hardwood Academic Pub., New York (1980)

[R11.76]  Walsh, V.C., Gram, H.: Classical and Neoclassical Theories of General Equilibrium: Historical Origins and Mathematical Structure. Oxford University Press, New York (1980)

[R11.77]  Weintraub, E.R.: General Equilibrium Analysis: Studies in Appraisal. Cambridge University Press, Cambridge (1985)

[R11.78]  Weintraub, E.R.: Microfoundations: The Compatibility of Microeconomics and Macroeconomics. Cambridge University Press, Cambridge (1980)

[R11.79]  Whittle, P.: Systems in Stochastic Equilibrium. Wiley, New York (1986)

## R.12.  Possible Worlds and the Knowledge Production Process

[R12.1]  Adams, R.M.: Theories of Actuality. Noûs 8, 211–231 (1974)

[R12.2]  Allen, S. (ed.): Possible Worlds in Humanities, Arts and Sciences. Proceedings of Nobel Symposium, vol. 65. Walter de Gruyter Pub., New York (1989)

[R12.3]   Armstrong, D.M.: A Combinatorial Theory of Possibility. Cambridge University Press (1989)

[R12.4]   Armstrong, D.M.: A World of States of Affairs. Cambridge University Press, Cambridge (1997)

[R12.5]   Bell, J.S.: Six Possible Worlds of Quantum Mechanics. In: Allen, S. (ed.) Proceedings of Nobel Symposium on Possible Worlds in Humanities, Arts and Sciences, vol. 65, pp. 359–373. Walter de Gruyter Pub., New York (1989)

[R12.6]   Bigelow, J.: Possible Worlds Foundations for Probability. Journal of Philosophical Logic 5, 299–320 (1976)

[R12.7]   Bradley, R., Swartz, N.: Possible World: An Introduction to Logic and its Philosophy. Bail Blackwell, Oxford (1997)

[R12.8]   Castañeda, H.-N.: Thinking and the Structure of the World. Philosophia 4, 3–40 (1974)

[R12.9]   Chihara, C.S.: The Worlds of Possibility: Modal Realism and the Semantics of Modal Logic. Clarendon (1998)

[R12.10]  Chisholm, R.: Identity through Possible Worlds: Some Questions. Noûs 1, 1–8 (1967); reprinted in Loux, The Possible and the Actual

[R12.11]  Divers, J.: Possible Worlds. Routledge, London (2002)

[R12.12]  Forrest, P.: Occam's Razor and Possible Worlds. Monist 65, 456–464 (1982)

[R12.13]  Forrest, P., Armstrong, D.M.: An Argument Against David Lewis' Theory of Possible Worlds. Australasian Journal of Philosophy 62, 164–168 (1984)

[R12.14]  Grim, P.: There is No Set of All Truths. Analysis 46, 186–191 (1986)

[R12.15]  Heller, M.: Five Layers of Interpretation for Possible Worlds. Philosophical Studies 90, 205–214 (1998)

[R12.16]  Herrick, P.: The Many Worlds of Logic. Oxford University Press, Oxford (1999)

[R12.17]  Krips, H.: Irreducible Probabilities and Indeterminism. Journal of Philosophical Logic 18, 155–172 (1989)

[R12.18]  Kuhn, T.S.: Possible Worlds in History of Science. In: Allen, S. (ed.) Possible Worlds in Humanities, Arts and Sciences, Proceedings of Nobel Symposium, vol. 65, pp. 9–41. Walter de Gruyter Pub., New York (1989)

[R12.19]  Kuratowski, K., Mostowski, A.: Set Theory: With an Introduction to Descriptive Set Theory. North-Holland, New York (1976)

[R12.20]  Lewis: David On the Plurality of Worlds. Basil Blackwell, Oxford (1986)

[R12.21]  Loux, M.J.: The Possible and the Actual: Readings in the Metaphysics of Modality. Cornell University Press, Ithaca & London (1979)

[R12.22]  Parsons, T.: Nonexistent Objects. Yale University Press, New Haven (1980)

[R12.23]  Perry, J.: From Worlds to Situations. Journal of Philosophical Logic 15, 83–107 (1986)

[R12.24]  Rescher, N., Brandom, R.: The Logic of Inconsistency: A Study in Non-Standard Possible-World Semantics And Ontology. Rowman and Littlefield (1979)

[R12.25]  Skyrms, B.: Possible Worlds, Physics and Metaphysics. Philosophical Studies 30, 323–332 (1976)

[R12.26]  Stalmaker, R.C.: Possible World. Noûs 10, 65–75 (1976)

[R12.27]  Quine, W.V.O.: Word and Object. M.I.T. Press (1960)

[R12.28]  Quine, W.V.O.: Ontological Relativity. Journal of Philosophy 65, 185–212 (1968)

# R13. Rationality, Information, Games, Conflicts and Exact Reasoning

[R13.1]   Aumann, R.: Correlated Equilibrium as an Expression of Bayesian Rationality. Econometrica 55, 1–18 (1987)

[R13.2]   Border, K.: Fixed Point Theorems with Applications to Economics and Game Theory. Cambridge University Press, Cambridge (1985)

[R13.3]   Brandenburger, A.: Knowledge and Equilibrium Games. Journal of Economic Perspectives 6, 83–102 (1992)

[R13.4]   Campbell, R., Sowden, L.: Paradoxes of Rationality and Cooperation: Prisoner's Dilemma and Newcomb's Problem. University of British Columbia Press, Vancouver (1985)

[R13.5]   Crawford, V., Sobel, J.: Strategic Information Transmission. Econometrica 50, 1431–1452 (1982)

[R13.6]   Scott, G., Humes, B.: Games, Information, and Politics: Applying Game Theoretic Models to Political Science. University of Michigan Press, Ann Arbor (1996)

[R13.7]   Gjesdal, F.: Information and Incentives: The Agency Information Problem. Review of Economic Studies 49, 373–390 (1982)

[R13.8]   Harsanyi, J.: Games with Incomplete Information Played by 'Bayesian' Players I: The Basic Model. Management Science 14, 159–182 (1967)

[R13.9]   Harsanyi, J.: Games with Incomplete Information Played by 'Bayesian' Players II: Bayesian Equilibrium Points. Management Science 14, 320–334 (1968)

[R13.10]  Harsanyi, J.: Games with Incomplete Information Played by 'Bayesian' Players III: The Basic Probability Distribution of the Game. Management Science 14, 486–502 (1968)

[R13.11]  Harsanyi, J.: Rational Behavior and Bargaining Equilibrium in Games and Social Situations. Cambridge University Press, New York (1977)

[R13.12]  Haussmann, U.G.: A Stochastic Maximum Principle for Optimal Control of Diffusions. Longman, Essex (1986)

[R13.13]  Istratescu, V.I.: Fixed Point Theory: An Introduction. Reidel Pub. Co., Dordrecht (1981)

[R13.14]  Krasovskii, N.N., Subbotin, A.I.: Game-theoretical Control Problems. Springer, New York (1988)

[R13.15]  Kuhn, H. (ed.): Classics in Game Theory. Princeton University Press, Princeton (1997)

[R13.16]  Lagunov, V.N.: Introduction to Differential Games and Control Theory. Heldermann Verlag, Berlin (1985)

[R13.17]  Luce, D.R., Raiffa, H.: Games and Decisions. John Wiley and Sons, New York (1957)

[R13.18]  Maynard Smith, J.: Evolution and the Theory of Games. Cambridge University Press, Cambridge (1982)

[R13.19]  Milgrom, P., Roberts, J.: Rationalizablility, Learning and Equilibrium in Game with Strategic Complementarities. Econometrica 58, 1255–1279 (1990)

[R13.20]  Myerson, R.: Game Theory: Analysis of Conflict. Harvard University Press, Cambridge (1991)

[R13.21]  Rapoport, A., Chammah, A.: Prisoner's Dilemma: A Study in Conflict and Cooperation. University of Michigan Press, Ann Arbor (1965)

[R13.22]  Roth, A.E.: The Economist as Engineer: Game Theory, Experimentation, and Computation as Tools for Design Economics. Econometrica 70, 1341–1378 (2002)

[R13.23]  Shubik, M.: Game Theory in the Social Sciences: Concepts and Solutions. MIT Press, Cambridge (1982)

[R13.24]  Smart, D.R.: Fixed point Theorems. Cambridge University Press, Cambridge (1980)

[R13.25]  Ulrich, H.: Fixed Point Theory of Parametrized Equivariant Maps. Springer, New York (1988)

[R13.26]  Von Neumann, J., Morgenstern, O.: The Theory of Games in Economic Behavior. John Wiley and Sons, New York (1944)

## R14. Rationality and Philosophy of Exact and Inexact Sciences in the Knoeledge Production

[R14.1]   Achinstein, P.: The Problem of Theoretical Terms. In: Brody, B.A. (ed.) Reading in the Philosophy of Science. Prentice Hall, Englewood Cliffs (1970)

[R14.2]   Amo Afer, A.G.: The Absence of Sensation and the Faculty of Sense in the Human Mind and Their Presence in our Organic and Living Body, Dissertation and Other essays 1727-1749. Jena, Martin Luther Universioty Translation, Halle Wittenberg (1968)

[R14.3]   Beeson, M.J.: Foundations of Constructive Mathematics. Springer, New York (1985)

[R14.4]   Benacerraf, P.: God, the Devil and Gödel. Monist 51, 9–32 (1967)

[R14.5]   Benecerraf, P., Putnam, H. (eds.): Philosophy of Mathematics: Selected Readings. Cambridge University Press, Cambridge (1983)

[R14.6]   Black, M.: The Nature of Mathematics. Adams and Co., Totowa (1965)

[R14.7]   Blanche, R.: Contemporary Science and Rationalism. Oliver and Boyd, Edinburgh (1968)

[R14.8]   Blanshard, B.: The Nature of Thought. Allen and Unwin, London (1939)

[R14.9]   Bloomfield, L.: Linguistic Aspects of Science. In: Neurath, O., et al. (eds.) International Encyclopedia of Unified Science, vol. 1-10, pp. 219–277. University of Chicago Press, Chicago (1955)

[R14.10]  Braithwaite, R.B.: Models in the empirical sciences. In: Brody, B.A. (ed.) Reading in the Philosophy of Science, pp. 268–275. Prentice-Hall, Englewood Cliffs (1970)

[R14.11]  Braithwaite, R.B.: Scientific Explanation. Cambridge University Press, Cambridge (1955)

[R14.12]  Brody, B.A. (ed.): Reading in the Philosophy of Science. Prentice-Hall, Englewood Cliffs (1970)

[R14.13]  Brody, B.A.: Confirmation and Explanation. In: Brody, B.A. (ed.) Reading in the Philosophy of Science, pp. 410–426. Prentice-Hall, Englewood Cliffs (1970)

[R14.14]  Brouwer, L.E.J.: Intuitionism and Formalism. Bull. of American Math. Soc. 20, 77–89 (1913); Also in Benecerraf, P., Putnam, H.(eds.) Philosophy of Mathematics: Selected Readings, pp. 77–89. Cambridge University Press, Cambridge (1983)

[R14.15]  Brouwer, L.E.J.: Consciousness, Philosophy, and Mathematics. In: Benecerraf, P., Putnam, H. (eds.) Philosophy of Mathematics: Selected Readings, pp. 90–96. Cambridge University Press, Cambridge (1983)

[R14.16]  Brouwer, L.E.J.: Collected Works. In: Heyting, A. (ed.) Philosophy and Foundations of Mathematics, vol. 1. Elsevier, New York (1975)

[R14.17]  Campbell, N.R.: What is Science? Dover, New York (1952)

[R14.18]  Carnap, R.: Foundations of Logic and Mathematics. In: International Encyclopedia of Unified Science, pp. 143–211. Univ. of Chicago, Chicago (1939)

[R14.19]  Carnap, R.: Statistical and Inductive Probability. In: Brody, B.A. (ed.) Reading in the Philosophy of Science, pp. 440–450. Prentice- Hall, Englewood Cliffs (1970)

[R14.20]  Carnap, R.: On Inductive Logic. Philosophy of Science 12, 72–97 (1945)

[R14.21]  Carnap, R.: The Two Concepts of Probability. Philosophy and Phenomenonological Review 5, 513–5532 (1945)

[R14.22]  Carnap, R.: The Methodological Character of Theoretical Concepts. In: Feigl, H., Scriven, M. (eds.) Minnesota Studies in the Philosophy of Science, vol. I, pp. 38–76 (1956)

[R14.23]  Charles, D., Lennon, K. (eds.): Reduction, Explanation, and Realism. Oxford University Press, Oxford (1992)

[R14.24]  Cohen, R.S., Wartofsky, M.W. (eds.): Methodological and Historical Essays in the Natural and Social Sciences. D. Reidel Publishing Co., Dordrecht (1974)

[R14.25]  Church, A.: An Unsolvable Problem of Elementary Number Theory. American Journal of Mathematics 58, 345–363 (1936)

[R14.26]  Dalen van, D. (ed.): Brouwer's Cambridge Lectures on Intuitionism. Cambridge University Press, Cambridge (1981)

[R14.27]  Davidson, D.: Truth and Meaning: Inquiries into Truth and Interpretation. Oxford University Press, Oxford (1984)

[R14.28]  Davis, M.: Computability and Unsolvability. McGraw-Hill, New York (1958)

[R14.29]  Dompere, K.K.: Polyrhythmicity: Foundations of African Philosophy. Adonis and Abbey Pub., London (2006)

[R14.30]  Dompere, K.K., Ejaz, M.: Epistemics of Development Economics: Toward a Methodological Critique and Unity. Greenwood Press, Westport (1995)

[R14.31]  Dummett, M.: The Philosophical Basis of Intuitionistic Logic. In: Benecerraf, P., Putnam, H. (eds.) Philosophy of Mathematics: Selected Readings, pp. 97–129. Cambridge University Press, Cambridge (1983)

[R14.32]  Feigl, H., Scriven, M. (eds.): Minnesota Studies in the Philosophy of Science, vol. I (1956)

[R14.33]  Feigl, H., Scriven, M. (eds.): Minnesota Studies in the Philosophy of Science, vol. II (1958)

[R14.34]  Frank, P.: Between Physics and Philosophy. Harvard University Press, Cambridge (1941)

[R14.35]  Garfinkel, A.: Forms of Explanation: Structures of Inquiry in Social Science. Yale University Press, New Haven (1981)

[R14.36]  Georgescu-Roegen, N.: Analytical Economics. Harvard University Press, Cambridge (1967)

[R14.37]  George, F.H.: Philosophical Foundations of Cybernetics. Tunbridge Well, Great Britain (1979)

[R14.38]  Gillam, B.: Geometrical Illusions. Scientific American, 102–111 (January 1980)

[R14.39]  Gödel, K.: What is Cantor's Continuum Problem? In: Benecerraf, P., Putnam, H. (eds.) Philosophy of Mathematics: Selected Readings, pp. 470–486. Cambridge University Press, Cambridge (1983)

[R14.40]  Gorsky, D.R.: Definition. Progress Publishers, Moscow (1974)

[R14.41]  Gray, W., Rizzo, N.D. (eds.): Unity Through Diversity. Gordon and Breach, New York (1973)

[R14.42]  Grattan-Guinness, I.: The Development of the Foundations of Mathematical Analysis From Euler to Riemann. MIT Press, Cambridge (1970)

[R14.43]  Hart, W.D. (ed.): The Philosophy of Mathematics. Oxford University Press, Oxford (1996)

[R14.44]  Hausman, D.M.: The Exact and Separate Science of Economics. Cambridge University Press, Cambridge (1992)

[R14.45]  Helmer, O., Oppenheim, P.: A Syntactical Definition of Probability and Degree of confirmation. The Journal of Symbolic Logic 10, 25–60 (1945)

[R14.46]  Helmer, O., Rescher, N.: On the Epistemology of the Inexact Sciences, P-1513. Rand Corporation, Santa Monica (October 13, 1958)

[R14.47]  Hempel, C.G.: Studies in the Logic of Confirmation. Mind, Part I 54, 1–26 (1945)

[R14.48]  Hempel, C.G.: The Theoretician's Dilemma. In: Feigl, H., Scriven, M. (eds.) Minnesota Studies in the Philosophy of Science, vol. II, pp. 37–98 (1958)

[R14.49]  Hempel, C.G.: Probabilistic Explanation. In: Brody, B.A. (ed.) Reading in the Philosophy of Science, pp. 28–38. Prentice- Hall, Englewood Cliffs (1970)

[R14.50]  Hempel, C.G., Oppenheim, P.: Studies in the Logic of Explanation. Philosophy of Science 15, 135–175 (1948); Also in Brody, B. A.(ed.): Reading in the Philosophy of Science, pp. 8–27. Prentice- Hall, Englewood Cliffs (1970)

[R14.51]  Hempel, C.G., Oppenheim, P.: A Definition of Degree of Confirmation. Philosophy of Science 12, 98–115 (1945)

[R14. 52]  Heyting, A.: Intuitionism: An Introduction. North-Holland, Amsterdam (1971)

[R14.53]  Hintikka, J. (ed.): The Philosophy of Mathematics. Oxford University Press, London (1969)

[R14.54]  Hockney, D., et al. (eds.): Contemporary Research in Philosophical Logic and Linguistic Semantics. Reidel Pub. Co., Dordrecht-Holland (1975)

[R14.55]  Hoyninggen-Huene, P., Wuketits, F.M. (eds.): Reductionism and Systems Theory in the Life Science:Some Problems and Perspectives. Kluwer Academic Pub., Dordrecht (1989)

[R14.56]  Kedrov, B.M.: Toward the Methodological Analysis of Scientific Discovery. Soviet Studies in Philosophy 1, 45–65 (1962)

[R14.57]  Kedrov, B.M.: On the Dialectics of Scientific Discovery. Soviet Studies in Philosophy 6, 16–27 (1967)

[R14.58]  Kemeny, J.G., Oppenheim, P.: On Reduction. In: Brody, B.A. (ed.) Reading in the Philosophy of Science, pp. 307–318. Prentice- Hall, Englewood Cliffs (1970)

[R14.59]  Klappholz, K.: Value Judgments of Economics. British Jour. of Philosophy 15, 97–114 (1964)

[R 14.60]  Kleene, S.C.: On the Interpretation of Intuitionistic Number Theory. Journal of Symbolic Logic 10, 109–124 (1945)

[R14.61]  Kmita, J.: The Methodology of Science as a Theoretical Discipline. Soviet Studies in Philosophy, 38–49 (Spring 1974)

[R14.62]  Krupp, S.R. (ed.): The Structure of Economic Science. Prentice-Hall, Englewood Cliffs (1966)

[R14.63]  Kuhn, T.: The Structure of Scientific Revolution. University of Chicago Press, Chicago (1970)

[R14.64]  Kuhn, T.: The Function of Dogma in Scientific Research. In: Brody, B.A. (ed.) Reading in the Philosophy of Science, pp. 356–374. Prentice-Hall, Englewood Cliffs (1970)

[R14.65]  Kuhn, T.: The Essential Tension:Selected Studies in Scientific Tradition and Change. University of Chicago Press, Chicago (1979)

[R14.66]  Lakatos, I. (ed.): The Problem of Inductive Logic. North-Holland, Amsterdam (1968)

[R14.67]  Lakatos, I.: Proofs and Refutations: The Logic of Mathematical Discovery. Cambridge University Press, Cambridge (1976)

[R14.68]  Lakatos, I.: Mathematics, Science and Epistemology: Philosophical Papers. In: Worrall, J., Currie, G. (eds.) vol. 2. Cambridge Univ. Press, Cambridge (1978)

[R14.69]  Lakatos, I.: The Methodology of Scientific Research Programmes, vol. 1. Cambridge University Press, New York (1978)

[R14.70]  Lakatos, I.: The Methodology of Scientific Research Programmes, vol. 1. Cambridge University Press, New York (1978)

[R14.71]  Lawson, T.: Economics and Reality. Routledge, New York (1977)

[R14.72]  Lenzen, V.: Procedures of Empirical Science. In: Neurath, O., et al. (eds.) International Encyclopedia of Unified Science, vol. 1-10, pp. 280–338. University of Chicago Press, Chicago (1955)

[R14.73]  Levi, I.: Must the Scientist make Value Judgments? In: Brody, B.A. (ed.) Reading in the Philosophy of Science, pp. 559–570. Prentice- Hall, Englewood Cliffs (1970)

[R14.74]  Lewis, D.: Convention: A Philosophical Study. Harvard University Press, Cambridge (1969)

[R14.75]  Mayer, T.: Truth versus Precision in Economics. Edward Elgar, London (1993)

[R14.76]  Menger, C.: Investigations into the Method of the Social Sciences with Special Reference to Economics. New York University Press, New York (1985)

[R14.77]  Mirowski, P. (ed.): The Reconstruction of Economic Theory. Kluwer Nijhoff, Boston (1986)

[R14.78]  Mueller, I.: Philosophy of Mathematics and Deductive Structure in Euclid's Elements. MIT Press, Cambridge (1981)

[R14.79]  Nagel, E.: Review: Karl Niebyl, Modern Mathematics and Some Problems of Quantity, Quality, and Motion in Economic Analysis. The Journal of Symbolic Logic, 74 (1940)

[R14.80]  Nagel, E., et al. (eds.): Logic, Methodology, and the Philosophy of Science. Stanford University Press, Stanford (1962)

[R14.81]  Narens, L.: A Theory of Belief for Scientific Refutations. Synthese 145, 397–423 (2005)

[R14.82]  Niebyl, K.H.: Modern Mathematics and Some Problems of Quantity, Quality and Motion in Economic Analysis. Philosophy of Science 7(1), 103–120 (1940)

[R14.83]  Neurath, O., et al. (eds.): International Encyclopedia of Unified Science, vol. 1 - 10. University of Chicago Press, Chicago (1955)

[R14.84]  Otto, N.: Unified Science as Encyclopedic. In: Neurath, O., et al. (eds.) International Encyclopedia of Unified Science, vol. 1-10, pp. 1–27. University of Chicago Press, Chicago (1955)

[R14.85]  Nkrumah, K.: Consciencism. Heinemann, London (1964)

[R14.86]  Planck, M.: The philosophy of Physics. Norton and Co., New York (1936)

[R14.87]  Planck, M.: Scientific Autobiography and Other Papers. Conn. Greenwood, Westport (1971)

[R14.88]  Planck, M.: The Meaning and Limits of Exact Science. In: Planck, M. (ed.) Scientific Autobiography and Other Papers, pp. 80–120. Conn. Greenwood, Westport (1971)

[R14.89]  Polanyi, M.: Genius in Science. In: Cohen, R.S., Wartofsky, M.W. (eds.) Methodological and Historical Essays in the Natural and Social Sciences, pp. 57–71. D. Reidel Publishing Co., Dordrecht (1974)

[R14.90]  Popper, K.: The Nature of Scientific Discovery. Harper and Row, New York (1968)

[R14.91]  Putnam, H.: Models and Reality. In: Benecerraf, P., Putnam, H. (eds.) Philosophy of Mathematics: Selected Readings, pp. 421–444. Cambridge University Press, Cambridge (1983)

[R14.92]  Reise, S.: The Universe of Meaning. The Philosophical Library, New York (1953)

[R14.93]  Robinson, R.: Definition. Clarendon Press, Oxford (1950)

[R14.94]  Rudner, R.: The Scientist qua Scientist Makes Value Judgments. Philosophy of Science 20, 1–6 (1953)

[R14.95]  Russell, B.: Our Knowledge of the External World. Norton, New York (1929)

[R14.96]  Russell, B.: Human Knowledge, Its Scope and Limits. Allen and Unwin, London (1948)

[R14.97]  Russell, B.: Logic and Knowledge: Essays 1901-1950, pp. 1901–1950. Capricorn Books, New York (1971)

[R14.98]  Russell, B.: An Inquiry into Meaning and Truth. Norton, New York (1940)

[R14.99]  Russell, B.: Introduction to Mathematical Philosophy. George Allen and Unwin, London (1919)

[R14.100] Russell, B.: The Problems of Philosophy. Oxford University Press, Oxford (1978)

[R14.101] Ruzavin, G.I.: On the Problem of the Interrelations of Modern Formal Logic and Mathematical Logic. Soviet Studies in Philosophy 3(1), 34–44 (1964)

[R14.102] Scriven, M.: Explanations, Predictions, and Laws. In: Brody, B.A. (ed.) Reading in the Philosophy of Science, pp. 88–104. Prentice- Hall, Englewood Cliffs (1970)

[R14.103] Sellars, W.: The Language of Theories. In: Brody, B.A. (ed.) Reading in the Philosophy of Science, pp. 343–353. Prentice- Hall, Englewood Cliffs (1970)

[R14.104] Sterman, J.: The Growth of Knowledge: Testing a Theory of Scientific Revolutions with a Formal Model. Technological Forecasting and Social Change 28, 93–122 (1995)

[R14.105] Tullock, G.: The Organization of Inquiry. Liberty Fund Inc., Indianapolis (1966)

[R14.106] Van Fraassen, B.: Introduction to Philosophy of Space and Time. Random House, New York (1970)

[R14.107] Veldman, W.: A Survey of Intuitionistic Descriptive Set Theory. In: Petkov, P.P. (ed.) Mathematical Logic: Proceedings of the Heyting Conference, pp. 155–174. Plenum Press, New York (1990)

[R14.108] Vetrov, A.A.: Mathematical Logic and Modern Formal Logic. Soviet Studies in Philosophy 3(1), 24–33 (1964)

[R14.109] von Mises, L.: Epistemological Problems in Economics. New York University Press, New York (1981)

[R14.110] Wang, H.: Reflections on Kurt. MIT Press, Cambridge (1987)

[R14.111] Watkins, J.W.N.: The Paradoxes of Confirmation. In: Brody, B.A. (ed.) Reading in the Philosophy of Science, pp. 433–438. Prentice- Hall, Englewood Cliffs (1970)

[R14.112] Whitehead, A.N.: Process and Reality. The Free Press, New York (1978)

[R14.113] Wittgenstein, L.: Ttactatus Logico-philosophicus. The Humanities Press Inc., Atlantic Highlands (1974)

[R14.114] Woodger, J.H.: The Axiomatic Method in Biology. Cambridge University Press, Cambridge (1937)

[R14.115] Zeman, J.: Information, Knowledge and Time. In: Kubát, L., Zeman, J. (eds.) Entropy and Information. Elsevier, Amsterdam (1975)

# R15. Riskiness, Decision-Choice Process and Paradoxes in Knowledge Constuction

[R15.1] Allais, M.: The Foundations of the Theory of Utility and Risk: Some Central Points of the Discussions at the Oslo Conference. In: Hagen, O., Wenstøp, F. (eds.) Progess in Utility and Risk Theory, pp. 3–131. D. Reidel Pub., Dordrecht (1984)

[R15.2] Allais, M., Hagen, O.: Expected Utility Hypotheses and the Allais Paradox. D. Reidel Pub., Dordrecht (1979)

[R15.3] Anand, P.: Foundations of Rational Choice Under Risk. Oxford University Press, New York (1993)

[R15.4] Anderson, N.H., Shanteau, J.C.: Information Integration in Risky Decision Making. Journal of Experimental Psychology 84, 441–451 (1970)

[R15.5] Bar-Hillel, M., Margalit, A.: Newcombe's Paradox Revisited. British Journal of Philosophy of Science 23, 295–304 (1972)

[R15.6] Campbell, R., Sowden, L. (eds.): Paradoxes of Rationality and Cooperation: Prisoner's Dilemma and Newcomb's Problem. Universith of British Columbia Press, Vancouver (1985)

[R15.7] Crouch, E.A., et al.: Risk/Analysis. Ballinger, Cambridge (1982)

[R15.8] Einhorn, H., Hogarth, R.M.: Ambiguity and Uncertainty in Probabilistic Inference. Psychological Review 92, 433–461 (1985)

[R15.9] Ellsberg, D.: Risk, Ambiguity and the Savage Axioms. Quarterly Journal of Economics 75, 643–669 (1961)

[R15.10] Friedman, M., Savage, L.J.: The Utility Analysis of Choice Involving Risk. Journal of Political Economy 56, 279–304

[R15.11] Peter, G., Sahlin, N.: Unreliable Probabilities, Risk taking, and Decision Making. Synthese 53, 361–386 (1982)

[R15.12] Handa, J.: Risk, Probability and a New Theory of Cardinal Utility. Journal of Political Economy 85, 97–122 (1977)

[R15.13] Harsanyi, J.C.: Cardinal Utility in Welfare Economics and in the Theory of Risk-Taking. Jour. Polit. Econ. 61, 434–435 (1953)

[R15.14]  Hart, A.G.: Risk, Uncertainty and Unprofitability of Compounding Probabilities. In: Lange, O., et al. (eds.) Mathematical Economics and Econometrics, pp. 110–118. Cicago University Press, Chicago (1942)

[R15.15]  Hurley, S.L.: Newcomb's Problem, Prisoner's Dilemma, and Collective Action. Synthese 86, 173–196 (1991)

[R15.16]  Kahneman, D., Tversky, A.: Prospect Theory. Econometrica 47, 263–292 (1979)

[R15.17]  Karmarkar, U.S.: Subjectively Weighted Utility and Allais Paradox. Organization Behavior and Human Performance 24, 67–72 (1979)

[R15.18]  Kogan, N., Wallach, M.A.: Risk Taking: A Study in Cognition and Personality. Hold Rinehart and Winston, New York (1974)

[R15.19]  Levi, I.: Ignorance, Probability and Rational Choice. Synthese 53, 387–417 (1982)

[R15.20]  Levi, I.: Four Types of Ignorance. Social Research 44, 745–756 (1977)

[R15.21]  MacCrimmon, Larsson, S.: Utility Theory: Axioms Versus 'Paradoxes'. In: Allais, M., Hagen, O. (eds.) Expected Utility Hypotheses and the Allais Paradox, pp. 333–409. D. Reidel Pub., Dordrecht (1979)

[R15.22]  Priest, G.: Sorites and Identity. Logique et Analyse 34, 293–296 (1991)

[R15.23]  Raiffa, H.: Risk, Ambiguity, and Savage Axioms:Comment. Quarterly Journal of Economics 77, 327–337 (1963)

[R15.24]  Roberts, H.V.: Risk, Ambiguity, and Savage Axioms:Comment. Quarterly Journal of Economics 75, 690–695 (1961)

[R15.25]  Sainsbury, R.M.: Paradoxes. Cambridge University Press, Cambridge (1995)

[R15.26]  Savage, L.J.: The Foundations of Statistics. Wiley, New York (1954)

[R15.27]  Shubik, M.: Information, Risk, Ignorance and Indeterminacy. Quarterly Journal of Economics 75, 643–669 (1961)

[R15.28]  Simpson, P.B.: Risk Allowance for Price Expectation. Econometrica 18, 253–259 (1950)

[R15.29]  Stigum, B.P., et al. (eds.): Foundations of Utility and Risk Theory with Applications. D. Reidel Publishing Com., Boston (1983)

[R15.30]  Williamson, T.: Knowledge and its Limits. Oxford University Press, Oxford (2000)

[R15.31]  Theil, H.: The Allocation of Power that Minimizes Tension. Operations Research 19, 977–982 (1971)

[R15.32]  Theil, H.: On Estimation of Relationships Involving Qualitative Variables. American Journal of Sociology 76, 103–154 (1970)

# R16.  The Prescriptive Science, Theory of Planning and Cost-Benefit Analysis in Knowledge Construction

[R16.1]  Alexander Ernest, R.: Approaches to Planning. Gordon and Breach, Philadelphia (1992)

[R16.2]  Bailey, J.: Social Theory for Planning. Routledge and Kegan Paul, London (1975)

[R16.3]  Burchell, R.W., Sternlieb, G. (eds.): Planning Theory in the 1980's: A Search for Future Directions. Rutgers University Center for Urban and Policy Research, New Brunswick (1978)

[R16.4]   Camhis, M.: Planning Theory and Philosophy. Tavistock Publication, London (1979)

[R16.5]   Chadwick, G.: A Systems View of Planning. Pergamon, Oxford (1971)

[R16.6]   Cooke, P.: Theories of Planning and Special Development. Hutchinson, London (1983)

[R16.7]   Dompere, K.K., Lawrence, T.: Planning. In: Hussain, S.B. (ed.) Encyclopedia of Capitalism, vol. II, pp. 649–653. Facts On File, Inc., New York (2004)

[R16.8]   Dompere, K.K.: Cost-Benefit Analysis and the Theory of Fuzzy Decision: Identification and Measurement Theory. Springer, Heidelberg (2004)

[R16.9]   Dompere, K.K.: Cost-Benefit Analysis and the Theory of Fuzzy Decision: Fuzzy Value Theory. Springer, Heidelberg (2004)

[R16.10]  Dompere, K.K.: Fuzziness and the Market Mockery of Democracy: The Political Economy of Rent-Seeking and Profit-Harvesting. A Working Monograph on political Economy, vol. II. Department of Economics, Howard University, Washington, DC

[R16.11]  Dompere, K.K.: Social Goal-Objective Formation, Democracy and National Interest: Political Economy under Fuzzy Rationality. A Working Monograph on Political Economy, vol. I. Department of Economics, Howard University, Washington, DC

[R16.12]  Faludi, A.: Planning Theory. Pergamon, Oxford (1973)

[R16.13]  Faludi, A. (ed.): A Reader in Planning Theory. Pergamon, Oxford (1973)

[R16.14]  Kickert, W.J.M.: Organization of Decision-Making A Systems-Theoretic Approach. North-Holland, New York (1980)

[R16.15]  Kickert, W.J.M.: Organization of Decision-Making A Systems-Theoretic Approach. North-Holland, New York (1980)

[R16.16]  Knight, F.H.: Risk, Uncertainty and Profit. University of Chicago Press, Chicago (1971)

[R16.17]  Knight, F.H.: On History and Method of Economics. University of Chicago Press, Chicago (1971)

## R17. Social Sciences, Mathematics and the Problems of Exact and Inexact Methods of Thought

[R17.1]   Ackoff, R.L.: Towards a System of Systems Concepts. Management Science 17(11), 661–671 (1971)

[R17.2]   Angyal, A.: The Structure of Wholes. Philosophy of Sciences 6(1), 23–37 (1939)

[R17.3]   Bahm, A.J.: Organicism: The Philosophy of Interdependence. International Philosophical Quarterly VII(2) (1967)

[R17.4]   Bealer, G.: Quality and Concept. Clarendon Press, Oxford (1982)

[R17.5]   Black, M.: Critical Thinking. Prentice-Hall, Englewood Cliffs (1952)

[R17.6]   Brewer, M.B., Collins, B.E.: Scientific Inquiry and Social Sciences. Jossey-Bass Pub., San Francisco (1981)

[R17.7]   Campbell, D.T.: On the Conflicts Between Biological and Social Evolution and Between Psychology and Moral Tradition. American Psychologist 30, 1103–1126 (1975)

[R17.8]   Churchman, C.W., Ratoosh, P. (eds.): Measurement: Definitions and Theories. John Wiley, New York (1959)

[R17.9]   Foley, D.: Problems versus Conflicts Economic Theory and Ideology. In: American Economic Association Papers and Proceedings, vol. 65, pp. 231–237 (1975)

[R17.10]  Garfinkel, A.: Forms of Explanation:Structures of Inquiry in Social Science. Yale University Press, New Haven, Conn. (1981)

[R17.11]  Georgescu-Roegen, N.: Analytical Economics. Harvard University Press, Cambridge (1967)

[R17.12]  Gilolispie, C.: The Edge of Objectivity. Princeton University Press, Princeton (1960)

[R17.13]  Hayek, F.A.: The Counter-Revolution of Science. Free Press of Glencoe Inc., New York (1952)

[R17.14]  Laudan, L.: Progress and Its Problems: Towards a Theory of Scientific Growth. University of California Press, Berkeley (1961)

[R17.15]  Marx, K.: The Poverty of Philosophy. International Pub., New York (1971)

[R17.16]  Phillips, D.C.: Holistic Thought in Social Sciences. Stanford University Press, Stanford (1976)

[R17.17]  Popper, K.: Objective Knowledge. Oxford University Press, Oxford (1972)

[R17.18]  Rashevsky, N.: Organismic Sets: Outline of a General Theory of Biological and Social Organism. General Systems XII, 21–28 (1967)

[R17.19]  Roberts, B., Holdren, B.: Theory of Social Process. Iowa University Press, Ames (1972)

[R17.20]  Rudner, R.S.: Philosophy of Social Sciences. Prentice-Hall, Englewood Cliffs (1966)

[R17.21]  Simon, H.A.: The Structure of Ill-Structured Problems. Artificial Intelligence 4, 181–201 (1973)

[R17.22]  Toulmin, S.: Foresight and understanding: An Enquiry into the Aims of Science. Harper and Row, New York (1961)

[R17.23]  Winch, P.: The Idea of a Social Science. Humanities Press, New York (1958)

## R18. Theories of Utility, Expected Utility and Exact Problems of Exact Methods

[R18.1]   Allais, M., Hagen, O. (eds.): Expected Utility Hypothesis and Allias Paradox. D. Reidel Pub., Dordrecht (1979)

[R18.2]   Chipman, J.S.: Foundations of Utility. Econometrica 28(2), 193–224 (1960)

[R18.3]   Eatwell, J., et al.: Utility and Probability. Norton, New York (1990)

[R18.4]   Fishburn, P.C.: The Foundations of Expected Utility. D. Reidel, Dordrecht (1982)

[R18.5]   Fishburn, P.C.: Utility Theory for Decision Making. Wiley, New York (1970)

[R18.6]   Luce, R.D., Suppes, P.: Preferences, Utility, and Subjective Probabilities. In: Luce, D.R., et al. (eds.) Handbook of Mathematical Psychology, pp. 42–49. Wiley, New York (1965)

[R18.7]   Mac Crimmon, K., Larsson, S.: Utility Theory Versus Paradoxes. In: Allais, M., Hagen, O. (eds.) Expected Utility Hypothesis and Allias Paradox. D. Reidel Pub., Dordrecht (1979)

[R18.8]   Samuelson, P.: Probability and Attempts to Measure Utility. Economic
          Review 1, 167–173 (1950)
[R18.9]   Samuelson, P.: Probability, Utility, and Independence Axiom. Econometrica 20,
          670–678 (1952)
[R18.10]  Schoemaker, P.: The Expected Utility Model: Its Variants, Purposes, Evidence
          and Limitations. Journal of Economic Literature 20, 529–563 (1982)
[R18.11]  Stigler, G.J.: The Development of Utility Theory I. Journal of Political
          Economy 58, 307–327 (1958)
[R18.12]  Stigler, G.J.: The Development of Utility Theory II. Journal of Political
          Economy 58, 373–396 (1958)
[R18.13]  Suppes, P., Winet, M.: An Axiomatization of Utility based on the Notion of
          Utility Differences. Management Science 1, 259–270 (1955)
[R18.14]  Von Mises, L.: The Ultimate Foundations of Economic Science. Sheed Andrews
          and McMeel, Kansas City (1962)
[R18.15]  von Neumann, J., Morgenstern, O.: Theory of Games and Economic Behavior.
          Princeton University Press, Princeton (1947)

# R19. Vagueness, Approximation and Reasoning in the Knowledge Construction

[R19.1]   Adams, E.W., Levine, H.F.: On the Uncertainties Transmitted from Premises to
          Conclusions in deductive Inferences. Synthese 30, 429–460 (1975)
[R19.2]   Arbib, M.A.: The Metaphorical Brain. McGraw-Hill, New York (1971)
[R19.3]   Becvar, J.: Notes on Vagueness and Mathematics. In: Skala, H.J., et al. (eds.)
          Aspects of Vagueness, pp. 1–11. D. Reidel Co., Dordrecht (1984)
[R19.4]   Black, M.: Vagueness: An Exercise in Logical Analysis. Philosophy of
          Science 17, 141–164 (1970)
[R19.5]   Black, M.: Reasoning with Loose Concepts. Dialogue 2, 1–12 (1973)
[R19.6]   Black, M.: Language and Philosophy. Cornell University Press, Ithaca (1949)
[R19.7]   Black, M.: The Analysis of Rules. In: Black, M. (ed.) Models and Metaphors:
          Studies in Language and Philosophy, pp. 95–139 (1962)
[R19.8]   Black, M.: Models and Metaphors: Studies in Language and Philosophy. Cornell
          University Press, NewYork (1962)
[R19.9]   Black, M.: Margins of Precision. Cornell University Press, Ithaca (1970)
[R19.10]  Boolos, G.S., Jeffrey, R.C.: Computability and Logic. Combridge University
          Press, New York (1989)
[R19.11]  Cohen, P.R.: Heuristic Reasoning about uncertainty: An Artificial Intelligent
          Approach. Pitman, Boston (1985)
[R19.12]  Darmstadter, H.: Better Theories. Philosophy of Science 42, 20–27 (1972)
[R19.13]  Davis, M.: Computability and Unsolvability. McGraw-Hill, New York (1958)
[R19.14]  Dummett, M.: Wang's Paradox. Synthese 30, 301–324 (1975)
[R19.15]  Dummett, M.: Truth and Other Enigmas. Harvard University Press, Cambridge
          (1978)
[R19.16]  Endicott, T.: Vagueness in the Law. Oxford University Press, Oxford (2000)
[R19.17]  Evans, G.: Can there be Vague Objects? Analysis 38, 208 (1978)
[R19.18]  Fine, K.: Vagueness, Truth and Logic. Synthese 54, 235–259 (1975)

[R19.19]  Gale, S.: Inexactness, Fuzzy Sets and the Foundation of Behavioral Geography. Geographical Analysis 4(4), 337–349 (1972)

[R19.20]  Ginsberg, M.L. (ed.): Readings in Non-monotonic Reason. Morgan Kaufman, Los Alamitos (1987)

[R19.21]  Goguen, J.A.: The Logic of Inexact Concepts. Synthese 19, 325–373 (1968)

[R19.22]  Grafe, W.: Differences in Individuation and Vagueness. In: Hartkamper, A., Schmidt, H.-J. (eds.) Structure and Approximation in Physical Theories, pp. 113–122. Plenum Press, New York (1981)

[R19.23]  Goguen, J.A.: The Logic of Inexact Concepts. Synthese 19 (1968-1969)

[R19.24]  Graff, D., Timothy (eds.): Vagueness. Ashgate Publishing, Aldershot (2002)

[R19.25]  Hartkämper, A., Schmidt, H.J. (eds.): Structure and Approximation in Physical Theories. Plenum Press, New York (1981)

[R19.26]  Hersh, H.M., et al.: A Fuzzy Set Approach to Modifiers and Vagueness in Natural Language. J. Experimental 105, 254–276 (1976)

[R19.27]  Hilpinen, R.: Approximate Truth and Truthlikeness. In: Prelecki, M., et al. (eds.) Formal Methods in the Methodology of Empirical Sciences, pp. 19–42. Reidel, Ossolineum, Dordrecht, Wroclaw (1976)

[R19.28]  Hockney, D., et al.: Contemporary Research in Philosophical Logic and Linguistic Semantics. Reidel Pub. Co., Dordrecht-Holland (1975)

[R19.29]  Ulrich, H., et al.: Non-Clasical Logics and their Applications to Fuzzy Subsets: A Handbook of the Mathematical Foundations of Fuzzy Set Theory. Kluwer, Boston (1995)

[R19.30]  Katz, M.: Inexact Geometry. Notre-Dame Journal of Formal Logic 21, 521–535 (1980)

[R19.31]  Katz, M.: Measures of Proximity and Dominance. In: Proceedings of the Second World Conference on Mathematics at the Service of Man, Universidad Politecnica de Las Palmas pp. 370–377 (1982)

[R19.32]  Katz, M.: The Logic of Approximation in Quantum Theory. Journal of Philosophical Logic 11, 215–228 (1982)

[R19.33]  Keefe, R.: Theories of Vagueness. Cambridge University Press, Cambridge (2000)

[R19.34]  Keefe, R., Smith, P. (eds.): Vagueness: A Reader. MIT Press, Cambridge (1996)

[R19.35]  Kling, R.: Fuzzy Planner: Reasoning with Inexact Concepts in a Procedural Problem-solving Language. Jour. Cybernetics 3, 1–16 (1973)

[R19.36]  Kruse, R.E., et al.: Uncertainty and Vagueness in Knowledge Based Systems: Numerical Methods. Springer, New York (1991)

[R19.37]  Ludwig, G.: Imprecision in Physics. In: Hartkämper, A., Schmidt, H.J. (eds.) Structure and Approximation in Physical Theories, pp. 7–19. Plenum Press, New York (1981)

[R19.38]  Kullback, S., Leibler, R.A.: Information and Sufficiency. Annals of Math. Statistics 22, 79–86 (1951)

[R19.39]  Lakoff, G.: Hedges: A Study in Meaning Criteria and Logic of Fuzzy Concepts. In: Hockney, D., et al. (eds.) Contemporary Research in Philosophical Logic and Linguistic Semantics, pp. 221–271. Reidel Pub. Co., Dordrecht-Holland (1975)

[R19.40]  Lakoff, G.: Hedges: A Study in Meaning Criteria and the Logic of Fuzzy Concepts. Jour. Philos. Logic 2, 458–508 (1973)

[R19.41]  Levi, I.: The Enterprise of Knowledge. MIT Press, Cambridge (1980)

[R19.42]  Łucasiewicz, J.: Selected Works. Studies in the Logical Foundations of Mathematics. North-Holland, Amsterdam (1970)

[R19.43] Machina, K.F.: Truth, Belief and Vagueness. Jour. Philos. Logic 5, 47–77 (1976)

[R19.44] Menges, G., et al.: On the Problem of Vagueness in the Social Sciences. In: Menges, G. (ed.) Information, Inference and Decision, pp. 51–61. D. Reidel Pub., Dordrecht (1974)

[R19.45] Merricks, T.: Varieties of Vagueness. Philosophy and Phenomenological Research 53, 145–157 (2001)

[R19.46] Mycielski, J.: On the Axiom of Determinateness. Fund. Mathematics 53, 205–224 (1964)

[R19.47] Mycielski, J.: On the Axiom of Determinateness II. Fund. Mathematics 59, 203–212 (1966)

[R19.48] Naess, A.: Towards a Theory of Interpretation and Preciseness. In: Linsky, L. (ed.) Semantics and the Philosophy of Language. Ill. Univ. of Illinois Press, Urbana (1951)

[R19.49] Narens, L.: The Theory of Belief. Journal of Mathematical Psychology 49, 1–31 (2003)

[R19.50] Narens, L.: A Theory of Belief for Scientific Refutations. Synthese 145, 397–423 (2005)

[R19.51] Netto, A.B.: Fuzzy Classes. Notices, Amar., Math. Society 28, 945 (1968)

[R19.52] Neurath, O., et al.: International Encyclopedia of Unified Science, vol. 1- 10. University of Chicago Press, Chicago (1955)

[R19.53] Niebyl, K.H.: Modern Mathematics and Some Problems of Quantity. Quality and Motion in Economic Analysis. Science 7(1), 103–120 (1940)

[R19.54] Orlowska, E.: Representation of Vague Information. Information Systems 13(2), 167–174 (1988)

[R19.55] Parrat, L.G.: Probability and Experimental Errors in Science. John Wiley and Sons, New York (1961)

[R19.56] Raffman, D.: Vagueness and Context-sensitivity. Philosophical Studies 81, 175–192 (1996)

[R19.57] Reiss, S.: The Universe of Meaning. The Philosophical Library, New York (1953)

[R19.58] Russell, B.: Vagueness. Australian Journal of Philosophy 1, 84–92 (1923)

[R19.59] Russell, B.: An Inquiry into Meaning and Truth. Norton, New York (1940)

[R19.60] Shapiro, S.: Vagueness in Context. Oxford University Press, Oxford (2006)

[R19.61] Skala, H.J.: Modelling Vagueness. In: Gupta, M.M., Sanchez, E. (eds.) Fuzzy Information and Decision Processes, pp. 101–109. North-Holland, Amsterdam (1982)

[R19.62] Skala, H.J., et al. (eds.): Aspects of Vagueness. D. Reidel Co., Dordrecht (1984)

[R19.63] Sorensen, R.: Vagueness and Contradiction. Oxford University Press, Oxford (2001)

[R19.64] Tamburrini, G., Termini, S.: Some Foundational Problems in Formalization of Vagueness. In: Gupta, M.M., et al. (eds.) Fuzzy Information and Decision Processes, pp. 161–166. North Holland, Amsterdam (1982)

[R19.65] Termini, S.: Aspects of Vagueness and Some Epistemological Problems Related to their Formalization. In: Skala, H.J., et al. (eds.) Aspects of Vagueness, pp. 205–230. D. Reidel Co., Dordrecht (1984)

[R19.66] Tikhonov, A.N., Arsenin, V.Y.: Solutions of Ill-Posed Problems. John Wiley and Sons, New York (1977)

[R19.67] Tversky, A., Kahneman, D.: Judgments under Uncertainty: Heuristics and Biases. Science 185, 1124–1131 (1974)

[R19.68] Ursul, A.D.: The Problem of the Objectivity of Information. In: Kubát, L., Zeman, J. (eds.) Entropy and Information, pp. 187–230. Elsevier, Amsterdam (1975)

[R19.69] Vardi, M. (ed.): Proceedings of Second Conference on Theoretical Aspects of Reasoning about Knowledge. Morgan Kaufman, Asiloman (1988)

[R19.70] Verma, R.R.: Vagueness and the Principle of the Excluded Middle. Mind 79, 66–77 (1970)

[R19.71] Vetrov, A.A.: Mathematical Logic and Modern Formal Logic. Soviet Studies in Philosophy 3(1), 24–33 (1964)

[R19.72] von Mises, R.: Probability, Statistics and Truth. Dover Pub., New York (1981)

[R19.73] Williamson, T.: Vagueness. Routledge, London (1994)

[R19.74] Wiredu, J.E.: Truth as a Logical Constant With an Application to the Principle of the Excluded Millde. Philos. Quart. 25, 305–317 (1975)

[R19.75] Wright, C.: On Coherence of Vague Predicates. Synthese 30, 325–365 (1975)

[R19.76] Wright, C.: The Epistemic Conception of Vagueness. Southern Journal of Philosophy 33(suppl.), 133–159 (1995)

[R19.77] Zadeh, L.A.: A Theory of Commonsense Knowledge. In: Skala, H.J., et al. (eds.) Aspects of Vagueness, pp. 257–295. D. Reidel Co., Dordrecht (1984)

[R19.78] Zadeh, L.A.: The Concept of Linguistic Variable and its Application to Approximate reasoning. Information Science 8, 199–249 (1975) (Also in vol. 9, pp. 40–80)

# Index